PLASTICS AND POLYMER
PROCESSING AUTOMATION

PLASTICS AND POLYMER PROCESSING AUTOMATION

Edited by

The Plastics and Rubber Institute

London, England

NOYES DATA CORPORATION

Park Ridge, New Jersey, U.S.A.

1987

Published in the United States of America by
Noyes Data Corporation
Mill Road, Park Ridge, New Jersey 07656

10 9 8 7 6 5 4 3 2 1

Library of Congress Cataloging-in-Publication Data

Plastics and polymer processing automation.

   Bibliography: p.
   Includes index.
   1. Plastics industry and trade--Automation--Congresses.
2. Chemical process control--Congresses. I. Plastics and
Rubber Institute.
TP1122.P555 1987   668.4   87-12401
ISBN 0-8155-1140-X

Foreword

This book is an authoritative, up-to-date discussion of advanced manufacturing technology – automation – in plastics and polymer processing, based on a conference sponsored by The Plastics and Rubber Institute. An amalgamation of the leading expertise from industry and academia is presented, which covers both technology and business strategy. The various chapters illustrate how to achieve increased competitiveness by properly planning the use of new technology.

In addition to ten chapters on the technology in general, specialized sections on computer aided/computer integrated manufacture and competitive manufacturing systems in continuous and discontinuous processes are also included. This book should be very helpful to those continually interested in cutting costs.

The information in the book is from *Polymer Processing: Automation '86*, prepared by The Plastics and Rubber Institute, based on The Plastic and Rubber Institute's second international conference on competitive manufacturing systems, held at Sandown Park, Surrey, UK, June 1986.

The table of contents is organized in such a way as to serve as a subject index and provides easy access to the information contained in the book.

NOTICE

The materials in the book were prepared for a conference sponsored by The Plastics and Rubber Institute. To the best of our knowledge the information in this publication is accurate; however the Publisher does not assume any responsibility or liability for the accuracy or completeness of, or consequences arising from, such information. Mention of company names or commercial products does not constitute endorsement or recommendation for use by the Publisher.

Plastics and polymer raw materials can be toxic, and therefore due caution should always be exercised in the use of these hazardous materials. Final determination of the suitability of any information, product, or equipment for use contemplated by any user, and the manner of that use, is the sole responsibility of the user. We strongly recommend that users seek and adhere to a manufacturer's or supplier's current instructions for handling each material or equipment they use.

Contents and Subject Index

**PART III
COMPETITIVE MANUFACTURING SYSTEMS IN
CONTINUOUS AND DISCONTINUOUS PROCESSES**

PART IV
SUMMATION

Part I

General

SUMMARY OF WELCOMING ADDRESS

V. John Osola*

Manufacturing has an important place in the economies of all
the developed countries - both in terms of employment and in its
contribution to wealth-generation. The rapid growth in the world-
wide application of microelectronics is providing new
opportunities for products, production processes, and manufacturing
control systems. It is also increasing competitive pressures.

Some three years ago the United Kingdom National Economic
Development Office commissioned a study of the impact of Advanced
Manufacturing Technology on a number of companies in the United
Kingdom, and the findings were published, and widely discussed in
the engineering industry during the second half of last year. The
study evaluated the AMT experience in those U.K. engineering
companies which had decided to upgrade their production processes,
and then to examine effects of a comprehensive AMT programme on
some typical performance factors. The report also reviewed the
costs and benefits of introducing AMT and gave guidance on
planning and implementation

The conclusions from this work were that :

1   successful investment in AMT provides substantial financial and
    organisational benefits to companies of all sizes;

2   the investment can be largely self-financing, with only a
    short-term requirement for external funding, provided that it
    is carried out in a number of well-considered and well-managed
    stages.

* Chairman, NEDO AMS Group Committee

The Advanced Manufacturing Systems Committee of NEDO believes that although the study was confined to batch engineering manufacturing operations, the principles and findings are generally applicable to a wide range of other manufacturing industries, and they will in due course be seeking firm data on this question.

For the engineering companies which were surveyed, the following key results are revealed :

1 Materials costs reduced by 13 - 15%

2 Total production costs reduced by 14 - 27%

3 Operating profits doubled or trebled

4 Reductions in staff numbers, floor space, and general overheads
   contribute to falls in unit costs and cost of sales.

Apart from profitability, other notable changes occur in fixed and working capital, cash flow, and in a shift in capital employed towards a higher fixed content. New technology allows working capital to be significantly reduced.

Success does not automatically follow investment in advanced manufacturing technology, however. Careful medium term planning is essential, coupled with detailed consultation with employees and trade unions. Investment in an employee training and re-training programme, at the same time as the capital investment programme takes place, is usually essential. Implementation in a number of discrete stages is advisable, achieving the full improvement and results from each stage before proceeding to the next. Discussions with engineering companies have suggested that there are about a dozen key steps involved in successful introduction of AMT and it is likely that similar steps will apply equally to other manufacturing companies in the process industries; the 12 key steps are :

1 A critical examination of the company's competitive performance

2 The establishment of a new manufacturing plan covering a
   3 - 5 year period

3 The appointment of suitably qualified manufacturing management

4 Detailed manufacturing planning based on proven technology and
   phazed in several manageable stages

5 Consultation with employees and trade unions

6 A detailed review of available skills within the company and
   the development of re-training programmes

7 The completion of full financial appraisal of the investment
   and its likely returns

8 The planning of stage I in fine detail

9 The implementation of stage I after financial approval received

10 The overhaul of stock records and production planning data

11 The improvement of manufacturing organisation and procedures
   coupled with a re-check of training requirements

12 The achievement of results from Stage I, the learning of
   lessons, and the detailed planning of Stage II based on Stage I
   experience.

Perhaps the most common mistake made by companies is to
concentrate on the selection of the most appropriate advanced
manufacturing systems and to pay far too little attention to the
people who are going to have to operate and manage those systems.
It cannot be emphasized too strongly that success will only be
achieved if equal attention is paid to the training, re-training,
management and motivation of the people directly concerned with
the new technology as is paid to the technology itself.

COMBINING JAPANESE METHODOLOGY WITH TECHNOLOGY TO ACHIEVE MORE COMPETITIVE MANUFACTURING SYSTEMS

J. Parnaby*

The problem of creating new competitive
manufacturing systems is addressed via a systems
engineering approach which integrates the
structured use of modern methodologies proven by
the Japanese, with relevant technologies. The
starting point is the question 'Why Innovate,
what's wrong with the way we have always
operated?'

A structured Manufacturing Systems Engineering
approach is described.

INTRODUCTION

Over the past 15 years a number of industry sectors have been
faced head-on with competition from Japan and many companies have
not survived. The Japanese success is not due to luck or
accident but rather it is a consequence of a totally professional
approach.

There is no single skill or technology which is responsible for
Japanese success, on the contrary Japanese Engineers and Managers
apply a total systems approach in which many elements are
consistently integrated and applied. They are true Manufacturing
Systems Engineers.

In contrast in Europe we have allowed the manufacturing function,
which often has the largest resource in a company, to become
uncompetitive. Often because of the type of people employed
in manufacturing engineering the manufacturing function has
had a technician culture in contrast to the graduate and
professional culture of other functions. We have tolerated
a panacea approach in which companies have desparately purchased
the lastest new technological aid in the hope that it would
work a miracle on a weak balance sheet. For example, a recent
survey (1) of nearly 200 installed F.M.S. high technology cells
showed that the average cell had a throughput of 8-12 components
per hour and manufactured a variety of 12 different types of
component. This is clearly neither economic or truly flexible.
Furthermore, many complex cells could not be effectively
maintained by operational staff and so had utilisations less
than 60%.

* Group Director, Manufacturing Technology, Lucas Industries

5

There is no substitute for a totally systematic and professional approach led by properly trained manufacturing systems engineers in which factory systems and organisation are methodically redesigned, with selective introduction of new technologies where appropriate. Companies must recognise that they have to compete through the development of a relevant competitive manufacturing strategy which blends selected technology with relevant methodologies in an integrated system.

WHY INNOVATE, WHAT'S THE PROBLEM?

Two ways to define our problem in the UK are given in Tables I and II below:

	Japan	Britain
Stock Turnover Ratio	20	4
Sales per Employee	£100k	£30k
Ratio of Indirect Staff to Directs	0.5	1.2
Product Cost	70%	100%
Manufacturing and Design lead-times	60%	100%
Delivery on time	95%	75%

Table I A Comparison of Typical British and Japanese Engineering Manufacturing Companies

Table II   Japanese Share of World Markets

35 mm Cameras	84%
Watches	82%
Motor Cycles	55%
Bicycles	12%
Telephones	66%
VCR's	84%
Colour Televisions	53%
Microwave Ovens	71%
Washing Machines	26%
Refrigerators	21%
Calculators	77%
Dry cells	31%

Such consistent performance cannot just be fortuitous

Thirty years ago Japan decided to compete in world markets via its manufacturing strategy. To break into these markets dominated by high volume manufacturers in Europe and the USA it had to find niches for products and so had to learn how to make high product varieties. Japanese manufacturers quickly discovered

that principles applied to low variety mass-production systems
in Europe and the USA could not be economic with high product
variety and created high stock costs and long lead times.

Also however, to encourage customers to buy, both price and
quality had to be competitive. Thus the skills developed by
Japan to achieve a competitive manufacturing strategy were

1 Economic manufacture of high product variety
2 Consistent high product quality at low cost
3 Flexible manufacturing systems and people systems

In contrast, in the UK as we saw Japanese competition develop
we tried to compete by incremental modification of existing
manufacturing systems which were often originally designed
as dedicated long changeover time production lines for very
low product variety and for efficient large batch manufacture
using highly specialised segmented labour.

Such piecemeal modification has resulted in overcomplex
manufacturing systems with overcomplex controls and a large
increase in indirect staff to handle the complexity. Further,
addition of complex computer systems has often worsened the
problem.

There is a need, therefore, to undertake fundamental re-design
of our manufacturing systems to create simpler quick changeover
flexible cellular units with simple control systems designed
specifically for the dynamic high product variety markets of
the 1980's. If we do this then we can compete via the correct
manufacturing strategy like the Japanese have done over the
past 20 years.

However, new flexible job functions are required in contrast
to the traditional highly fragmented structures.

IS THERE A STRUCTURED APPROACH?

As a consequence of the technician culture in manufacturing and
the neglect of professional manufacturing development over the
past 30 years, the design of competitive manufacturing systems
has become perceived to be a black art. However, there is a
structured manufacturing systems engineering approach to the
systematic design of competitive manufacturing systems and the
achievement of simplicity. This approach can be described briefly
in the following phases:

Phase 1 Setting Relevant Targets:
1   Competitor Analysis and best practice survey
2   Define quantified target business ratios as follows:
         Ratio of indirects and staff to directs
         Stock Turnover Ratio
         Value Added per Employee
         Manufacturing Lead Times
         Percentage reduction in product cost
         Percentage reduction in quality rejects
         Percentage reduction in no-value added activities in
            factory flow charts

Phase 2   Achieving the Targets:
1   Train all management from the top in the principles of modern
        manufacturing systems engineering including Japanese
        systems methodologies
2   Recruit a top flight Manufacturing Development Manager with
        manufacturing systems engineering skills
3   Set up a development Task Force and train them in
        manufacturing systems engineering skills
4   Set up project communication procedures
5   Initiate the manufacturing, business and organisation redesign
        project to achieve the targets set above and review
        progress by the General Manager every month with insistence
        on achievement of targets
6   Create a new cellular manufacturing system by rearrangement of
        facilities with new flexible jobs and train all personnel
        before putting them into the new jobs
7   Audit performance achievement against design performance at
        each stage of implementation
8   Initiate a supplier development programme

The systematic manufacturing system design process can be defined
in the following five stages: (2)

Stage 1   –   Data collection on markets, products, process, flow-
              charts and machine capabilities. Search for patterns
              and natural cellular architectures

Stage 2   –   Steady state design based on subsystem input-output
              analysis to define basic configurations of cells,
              machines, processes and people and their matching
              interface requirements

Stage 3   –   Dynamic design to account for areas of potential
              change, e.g. in product mix

Stage 4   –   Information flow system and database definitions

Stage 5   –   Control function definition, control system design
              and overall control systems integration for day to day
              operations management

It is essential that such an approach is well understood by the
development teams required to re-design many of the current
uncompetitive British manufacturing systems to guide them through
the complex processes. Also it is most important that such systems
engineering methodologies are taught widely to undergraduate
engineers.

WHAT IS THE ROLE OF EDUCATION AND TRAINING?

The creation of new manufacturing systems has to be a knowledge-
led process. Traditional specialised Production Engineering has
failed to meet the Japanese competitiveness challenge.

Our educational institutions have consistently produced over-
specialised science oriented engineers and are generally not up-
to-date in the teaching of modern manufacturing and systems
engineering principles (3).

This leads to two requirements:

1   Industry must continually press our academic institutions to change from their traditional ways and provide more courses to train Manufacturing Systems Engineers and to understand all aspects of Manufacturing Systems as defined below: (4)

```
 ___________________________________________________________________
|                                                                   |
|              DEFINITION OF A MANUFACTURING SYSTEM                 |
|                                                                   |
| An integrated combination of processes, machine systems,          |
| people, organisational structures, information flows, control     |
| systems and computers whose purpose is to achieve economic        |
| product manufacture and internationally competitive               |
| performance.  The system has defined but progressively            |
| changing objectives to meet, some of which can be quantified      |
| and others, such as those relating to responsiveness,             |
| flexibility and quality of service, whilst being extremely        |
| important are difficult to quantify.  Nevertheless, the system    |
| must have integrated controls which systematically operate it     |
| to ensure that the competitiveness objectives are continually     |
| met and which adapt to change.                                    |
|                                                                   |
|___________________________________________________________________|
```

2   In the short term industry must invest in the development of internal continuing education courses to quickly update existing managers and engineers in the principles of modern manufacturing systems design, operation and control. In doing so it has to be noted that it will be difficult to find external help from academic institutions since most academic engineers do not understand manufacturing systems engineering. Also the cost of this type of development is high, particularly if all levels of staff are to be trained from General Manager down to Operators. In Lucas Industries in 1985 the cost of in-house training totalled £45m out of a sales turnover of £1.5bn.

WHAT IS JAPANESE METHODOLOGY?

The Japanese are the world's best manufacturing systems engineers and their impressive manufacturing performance is not simply the result of investment in robots, automation and computers. Similarly, the high level of unemployment in the UK is not due to automation robots or computers. We have lost market share because our manufacturing systems designed using 1950's production engineering principles are not competitive. Other people, including the Japanese, are making our products. Japanese Engineers design simple manufacturing systems with simple controls (5) **and create small cellular manageable units inside large units.** The basic elements of Japanese Systems Engineering methodology are shown in Figure I. Most of the systematic and simple principles they use have been adapted from USA and European practice. Figure 2 indicates the principle of selective integration of new technology with Japanese methodology.

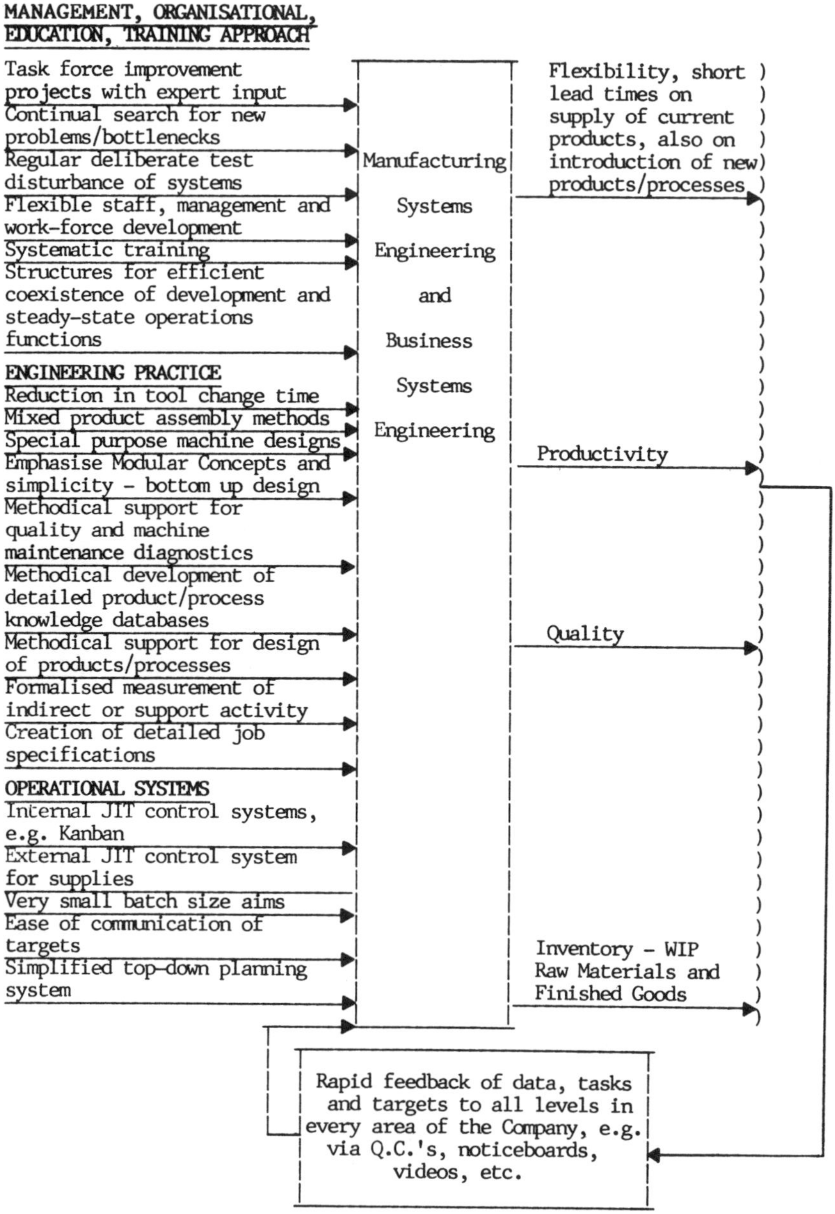

Figure I   JAPANESE SYSTEMS APPROACH TO PRODUCTIVITY, INVENTORY REDUCTION AND QUALITY CONTROL

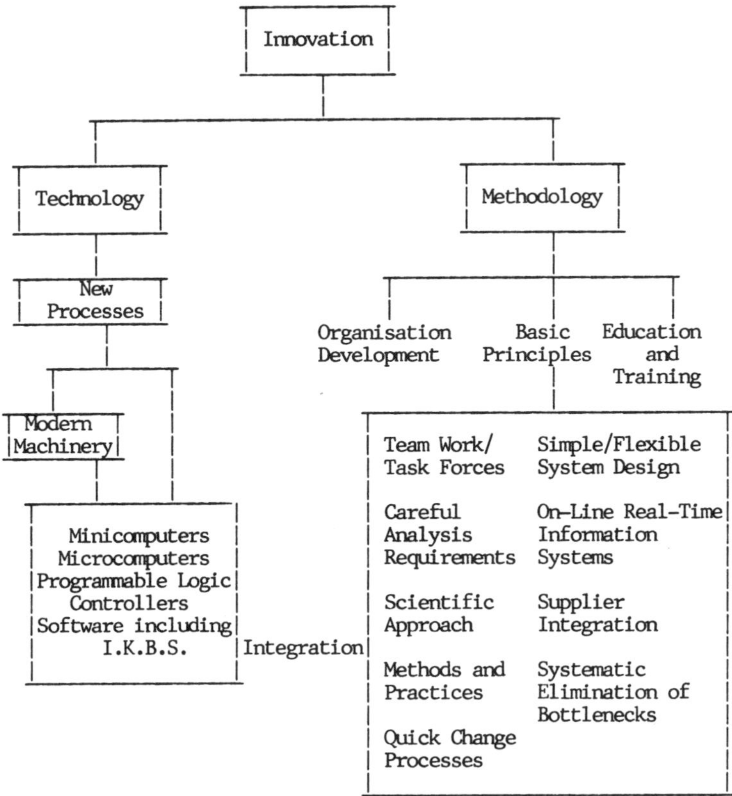

Figure II   INNOVATION AND IMPROVING MANUFACTURING SYSTEMS

CAN BRITAIN COMPETE WITH JAPANESE MANUFACTURING INDUSTRY?

There is increasing evidence in USA, Britain and Europe that companies who make the effort to learn new systems engineering skills and create fundamental new types of factories to match 1980's market needs, can compete with Japan. (6)

Two examples from Lucas Industries where the approach described in this paper have been applied are:

1   Automotive Electronic equipment factory over 1 year:

  Stock reduction £3m
  Manufacturing lead time reduction by 10
  Productivity increase by 35%
  Quality reject reduction by 5 times
  J.I.T. Kanban system introduced and piecework removed

2   Automotive Electrical equipment factory over 2 years:

  Stock turnover ratio increased from 7 to 13 times
  Productivity increase 25%
  Lead times reduced by a factor of 5
  Kanban systems introduced, piecework removed
  Production control and progress chasing department
   eliminated

Whilst there is much more work to be done, the progress described above is most encouraging.

REFERENCES

1   Flexible Manufacturing Systems – Myth and Reality. UWIST paper in Int. Jour. of Adv. Manufacturing Systems, 1985

2   The Design of Competitive Manufacturing Systems, J Parnaby, Int. J. Technology Management, Vol. 1, No. 1, 1986

3   Education and Training in Manufacturing Systems Engineering, J Parnaby, I.Prod.E., PEP '86 Conference, July 1986

4   Computers in Engineering and Manufacture, J Parnaby, N D Burns, M F Hessey, A Larner, S K Bhattacharyya, Collins Publishers, to be published September 1986

5   Study of Toyota Production System from Industrial Engineering viewpoint. S Shingo, Japan Management Association, Published by Productivity Inc., P O Box 814, Cambridge, MA 02238, USA

6   S Walleck, Director, McKinsey & Co., Chief Executive Journal, Winter 85/86, pp. 12–16.

THE LINK BETWEEN ADVANCED MANUFACTURING TECHNOLOGY AND COMPETITIVE STRATEGY IN COMMERCIAL POLYMER PROCESSING SYSTEMS

M.R. Tee*

ABSTRACT

A relationship between new technology and competition is intuitively acceptable but allocating informal status to such recognition can cause technology to be excessively decoupled from the search for competitive advantage. Reviews of Advanced Manufacturing Technology (AMT) coverage and a classical approach to competitive strategy formulation suggest that a more formal approach, capable of including the wider financial issues involved in AMT project implementation, is possible.

Prefacing Commentary

To do full justice to any study of the relationship between the three centres of interest in the title i.e. Advanced Manufacturing Technology, Competitive Strategy and Commercial Polymer Processing Systems, would require a volume of significant proportions to be written and this cannot be our intent here. For this reason the list of references supplied is reasonably extensive but to keep such a list to sensible proportions, a great deal of culling has unfortunately been found necessary.

In addition, the great speed with which new technology is becoming available for use - in a wide variety of applications on an international basis, means that the initial research required by this paper will inevitably have missed some significant publications. It has therefore to be hoped that the linkage achieved here does justice to the quality of the original work contained in the references and the importance of the subject matter.

* Knight Wendling Ltd

Summary of Conclusions

In the following we invoke three main propositions, move on to accept that significant extenuating circumstances, having the ability to constrain progress in the polymer processing sector exist, and then derive linkages between Advanced Manufacturing Technology (AMT) and competitive strategy. The scope of the official DTI definition of AMT has however had to be extended during such derivation.

The derived linkages, avoiding intuitive acceptance of the fact that a link will exist, are based on that which is referred to here as the Porter Model of the five forces which drive competition, the three generic strategies formulated by the same author and the well known phenomenon, the experience curve.

The propositions are:

1)   That the full extent of technology available to decision makers in the polymer processing sector is not in general well understood and furthermore that its relationship to the operations of companies of varying size and differing culture is even less well understood in strategic terms.

2)   That competitive strategy formulation is not something that is included in a formalised and structured approach to decision-making in the sector, possibly because of the demotivating volume of analysis required.

3)   That any link between technology generally and competitive strategy is only informally recognised, in spite of a very large volume of historical evidence which suggests a very positive link.

The implications are that the level of technology available to the polymer processing sector is in excess of that which it is apparently capable of absorbing and that technology in general is excessively decoupled from competitive strategy formulation.

In consequence it is implied that at a time when the severity of competition generally is expected to increase, manufacturing organisations in the sector are actually sub-optimising, or at risk of sub-optimisation, in terms of their ability to compete through manufacturing.

To retain a link with a possible sector view of reality and hence overall credulity, it needs to be accepted that:

1)   Real engineering constraints act ·to limit the use of computers in polymer processing, in the area of process control.

2)   The industrial sector in which an attempt is being made to link two powerful, if relatively unexplored fields, possesses a very high level of diversity and, in consequence, becomes one that is difficult to analyse.

3)   An important element of diversity relates to company size and hence company resources in both human and financial terms.

4)   The investment programmes required have a highly visible and demotiva-
     ting work load, particularly at the planning and implementation stage
     and are accompanied by a demand for scarce skills.

5)   All of the above factors impact on an industry sector not over-endowed
     with retained profits available for re-investment.

Some decision makers in the sector are not convinced that claims for the
returns likely to become available from investment in AMT will be achieved
and some sympathy with scepticism exists in the literature(1)(2) especially
when high technology projects are considered. The fact that some companies
may have become engaged on under planned schemes with little practical
experience is rarely mentioned and few, if any, post-expenditure audits
from the sector are published. In other sectors however and particularly
in the US, audits of AMT implementation projects have been carried out and
the indication is that positive improvements in AMT project management have
occured over the last 3 years. Some implementation guidelines are
fortunately now available which are applicable to projects of all sizes.

Activity directed towards the improvement of control over the internal and
external dynamics involved in the operation of a business, using the
generation and application of better information will always be difficult.
However, as the previous paper will have shown - it needs to be done and
the following paper provides a view on how it might be done.

When considering investment appraisal techniques it is suggested that sole
consideration of the risk free rate of return on capital will be found to
be inadequate. This will be particularly so if the domestic or foreign
competition includes forecasts of structural change in the industry within
its strategic reasoning. It would appear prudent to anticipate an increasing volume of international trade in the products of the sector and this
can only make the problem of strategic analysis referred to earlier, even
more difficult.

Whilst the smaller company has to be encouraged by the contents of the
Osola Committee Report and reports of the advantages to be gained at low
cost from initial improvements to manufacturing systems, it needs also to
be aware of the implicit threat in any recognised abilities of AMT to
emphasise economies of scale. This is inevitable, as the boundaries of
positioning policy become less distinct in high level manufacturing system
design. Whether a point exists at which economies of scale move into the
area of diminishing returns is debatable. Having the ability to address
the question from an informed viewpoint, is however important for every
sector company.

The objective of all competitive strategy is to secure a position of maximum defensibility long term. Given that a link between AMT and competitive
strategy is recognised, the company-specific threats and opportunities
represented by the former need to be well researched and cost/benefit
analyses effectively carried out. This requires use of a logic that is
forward looking and capable of challenging existing company culture.

Unsuccessful competitors will be found amongst the unwary and probably also amongst under-resourced companies if our own sector cannot fully agree the Osola Committee findings. Companies that learn to adapt to an environment made subject to increased levels of change by AMT, whether they be big or small, will survive this new initiator of evolutionary change but the essential pre-requisite is adequate knowledge.

Through conferences such as this, its predecessor and successors, it is hoped that some of the inputs that will assist survival will be gained.

From the competitive strategy viewpoint two essential inputs exist. The first, INTELLIGENCE, requires a company to know what is actually happening in technical and commercial terms in the market place on which it depends for its survival. The second, FORECASTING SKILL, allows the prediction of what is expected to happen and provides the single most useful and important database for operations management.

Thereafter, it will be the use of human skills - directed towards best use of the total resources available within the techno-social combination that is a manufacturing organisation, which will determine the degree of success achieved.

In view of a possible increase in the rate of change in the sector it is suggested that if it has not already been done, each Chief Executive would draw benefit from a serious and purposeful audit of his organisations asset base in terms of:

*   the knowledge that it requires for long-term security and the knowledge it holds.

*   the physical resources it requires for long-term security and the physical resources it owns or over which it has long-term control.

*   the human resources it requires for long-term security and the human resources that currently contribute to company performance.

*   the financial resources it requires for long term security and the financial resources upon which it can now call.

Long-term in this instance means a period of at least 3-5 years.

If an explicit competitive strategy does not exist, it is suggested that the above will assist the first steps towards the achievement of definition.

When assessing any linkages between competitive strategy and AMT it is clearly necessary to know what AMT has to offer, both to the company concerned and its competitors, with the latter assessment being considered to be more important. It needs also to be recognised that some of the techniques available under an extended AMT definition can assist such assessments considerably.

Whilst potential linkages between AMT and competitive strategy can be found by resort to existing theory the most important single link is probably the decision maker that knows fully what is on offer, what might become on offer within a planning period that will preferably extend to 5 years and can include both in the detailed planning that is fundamental to the use of AMT in the search for advantage.

To date, four stages have been recognised(3) which can form the basis of progression towards manufacturing-led business advantage and clearly the use of AMT has a central part to play in each.

With our concern for our manufacturing base in the UK and in 'Industry Year', the time to take stock and revamp manufacturing strategy might well be now, and hopefully the follow papers will help to strengthen the arm.

General Discussion

Kantrow(4), in the late Summer of 1980, advised that the past decade had revealed a growing awareness amongst managers of the need to incorporate technological issues within strategic decision making. The increasing discovery that technology and strategy are inseparable is reported and the review of the research he provided is especially helpful.

With hindsight one might wonder why such 'discoveries' had taken so long, but when initially considering the above title, observation unavoidably led to the belief that within the polymer processing sector, one is attempting to link two fields, each incompletely explored in conceptual and practical terms by the majority of sector company decision-makers.

To offset any sense of surprise arising out of the above, one needs to be reminded that the normality of the technology/strategy knowledge gap would almost certainly be validated by a serious historical review of the London Share Service information in the 'Financial Times'.

With thought and hindsight, the following quotation from Schumpeter(5) is all too acceptable.

> 'Technological Change - The source of creative destruction by which monopolies are destroyed and new industries created.'

By concentrating here on manufacturing technology it is accepted that a less global view of technology than that assumed by Kantrow is being taken, but the lack of exploration referred to is thought likely to be common to both global and more highly focussed analyses.

When referring to manufacturing considerations separately, Kantrow regards any decoupling of the technological dimension of manufacturing operations from overall strategic thinking as sheer foolishness. The results of further research papers, which are well-referenced in his own publication, are marshalled to support the view that operations management and the management of technical innovation and change are inextricably linked. Furthermore, it is postulated that manufacturing operations and technology, properly understood, are inextricably linked to the full range of strategy-related considerations.

Clearly, the inclusion of the qualification 'properly understood' is important and 'understanding' is something that one hopes will be assisted by a sector conference of this kind.

In a return to specifics, some backing for the above stated view of incomplete exploration in the polymer processing sector - is provided by the results of a survey carried out by the Plastics Industry Training Board in 1984(6). This is supplemented by work supported jointly by the Science and Engineering Research Council (SERC/PED) and The British Plastics Federation (BPF)(7).

Figure 1 is extracted from the PED/BPF study and is also referred to in a review by Coates(8). In the latter it is stated that the principal reasons, managerial factors excluded, underlying the relatively low use of computers in polymer processing are felt to be:

*   The existence of non-standard applications

*   The basic lack of detailed knowledge relating to complex multi-variable operations such as polymer processing.

Whilst, for very good engineering reasons, it is doubted whether over the last 2-3 years the rate of progress in the above two areas will have been rapid, it is also doubted whether progress in areas not limited by engineering constraints give cause for satisfaction. To have achieved any sense of satisfaction it would have been required to prove that adequate notice has been taken of a major report(9) covering Manufacturing Industry and New Technology, from which the following extracts have been taken.

'Manufacturing industry is vital to the UK but is under considerable pressure to increase its competitiveness. Companies using modern manufacturing technologies effectively are able to achieve higher productivity and efficiencies.

New and advanced manufacturing technologies now being developed offer even greater scope for improved productivity and product quality. Those firms which do not make use of these technologies risk being overtaken by competitors achieving superior quality at lower cost.'

In the above quotation, a link between Advanced Manufacturing Technology (AMT) and competitive strategy is being formed, albeit in a somewhat limited manner but additional emphasis and sense of timing is gained using a second quote from the same report.

'Too many companies have not yet applied advanced manufacturing technology to their manufacturing process. Those companies will become progressively less competitive and many will not survive the next ten years. Now is the time to invest. Many forms of finance are available from a wide variety of sources. In our view, the principal cause of failure to invest is a lack of conviction of the vital importance of doing so.'

It is significant to note that the above comments would have been made in the full knowledge that, in the five years leading up to the date of publication of the ACARD report, advances in the application of technology to the manufacturing area in the developed countries had easily outstripped those made in the last two decades.

Many present will undoubtedly recognise the latent exhortation for a quantum leap in investment and many might feel that the explicit reference to the availability of finance is excessively overgeneralised and not particularly helpful. However, the latent message is that any absence of investment requires a conscious decision not to invest and such a decision needs to be fully defensible in any open and well-reasoned debate. The implicit argument is that it is doubted whether a sound case for non-investment can be made.

It would appear that no additional quantitative data has been added to the original PPITB report and in consequence the plastics processing industry can be assumed to remain a sector where awareness of new technology, attitude and conviction can improve with company size. It is also an industry where relative company size distribution possibly acts to place it behind the manufacturing engineering sector in an all round approach to AMT.

There is no reason to suppose that the situation with rubber processing is any way different to plastics processing although it is unlikely that it is in any way further advanced.

The PPITB report does however acknowledge that:

> 'In every process in the Plastics Industry and in every size sector, there can be found an example of successful and comprehensive use of new technology.'

The saying 'He that seeks finds' would still therefore appear valid but for whatever reason the Industry would seem to lack seekers.

If this is true for present day technology, it is observed to be especially true for the technology that is as yet on the horizon. Few senior managements in the sector appear to have been able, for whatever reason, to form any opinion as to the real strategic value of computer usage to their operations, particularly when the hardware is placed in a modelling or forecasting mode. Yet the ability to predict the likely impact and timing of any form of innovation, once the nature of such innovation is known, clearly aids the search for competitive advantage.

In a sector where 80% of sector companies reportedly(10)(11) employ less than 100 people the existence of a technology/strategy knowledge gap is not difficult to understand. For many, the major projects which hit the technical press are of passing interest and the reported (1),(2),(12) inability of most of the earlier FMS introductions to realise forecast expectations will not have gone unnoticed. Dempsey, Foyer and Walleck have however taken a very positive approach to rectification in their papers and it could well be that further commentary will be provided in the following presentation.

Those that lack conviction are however directed towards the findings of the Osola Committee(13) and a recent paper in the PRODUCTION ENGINEER(14). In the latter, an overview of FMS and CIM progress in Japan, anticipated developments of FMS were indicated in the form shown in Figure 2. The points of interest are the activity arms which show anticipated movements towards:

*   the development of compact and low cost FMS Systems.

*   the expansion of FMS to fields other than machining manufacturing

    *   the development of higher level systems.

It is known that several companies in the UK polymer processing sector are developing manufacturing system designs which can be described as compact flexible manufacturing systems and in consequence are already involved in one or more of the tendencies of recently developed FMS designs in Japan referred to by Professor IWATA.

i.e. *   Enforcement of completely unattended operations during night shifts due to improvement of system reliability.

    *   Development of automatic work handling devices

    *   Enrichment of software in the particular application of dynamic scheduling.

    *   Development of low cost Flexible Manufacturing Cells (FMC)

In any audits directed towards definition of a firm's position within the sector, these tendencies and others which are not difficult to perceive, will need to be given serious consideration.

It is emphasised that the main purpose of this paper is not to make a case for or against investment. It is, however, very necessary to make the point when considering any link between AMT and competitive strategy that with the present high rate of change, strong conviction - one way or another, in regard to investment is required. It will not be found sufficient to be made aware from conferences such as this of the AMT experience in our own or related sectors, and do nothing.

In doing something, activity based on inputs which promote proactive rather than reactive responses are clearly preferable as these are more likely to be linked to strategic reasoning.

Fortunately for the sector, the supply side of the industry, very ably supported by the trade and professional press has provided an excellent service in supplying much of the information required and very excellent references abound. There is, however, little the supply side can do in terms of generating an actual commitment to the avoidance of drift by the decision makers. The latter need, for their own protection, to ensure that such references do not remain buried in the pile of reading matter that pricks all well developed consciences but never seems to get any smaller.

It is reasonable to suppose that the level of investment in new technology will remain generally linked to company size, although in strategic terms this will require subsidiary company size to be linked to ultimate holding company size.

Given that a common need exists for all companies to avoid investment in new technology for technology's sake, the general highly simplified relationship shown in Figure 3 is suggested likely to exist. The degree of scatter experienced might be expected to be large and in an ideal sense directly associated with a company's level of understanding of and commitment to a defined competitive strategy. In every case this will be company -specific.

It is with the ability to define the level of investment required over a given period of time, following an assessment of strategic requirements, that the link between technology and competitive strategy is most usefully and precisely formed. That it seldom seems to happen within the sector is thought to be associated with maturity factors and the lack of consideration that technological potential often receives from top management teams. It seems reasonable to suppose however that during any time of rapid technological change, an explicit competitive strategy able to define and marshall the resources needed to introduce a high quality productive system on positive cost/benefit terms, will be able to exploit to the full the opportunity that such a period of rapid technological change presents.

At this stage it is possible to suggest that economy of scale considerations may be prompted by computerisation studies and the management of the smaller undifferentiated company is seen likely to become more difficult if the investment stakes are raised. It is also known that many manufacturing facilities characterised by multiple assembly activity i.e. Automotive Manufacture, Consumer Durables Manufacture etc., are aware of the potential offered by higher level systems - and the demands that their eventual introduction they might place for an increased volume of on-site or near-site component manufacture. It is easy to envisage a possible reduction of BOF components from the polymer processing sector in a high level system, particularly by companies already in possession of extensive polymer processing expertise. Within the component supply sector it is therefore conceivable that the above will act to increase the number of unsuccessful competitors. This may well give rise to significant structural change as successful competitors move purposefully to achieve the positions most easily defended over the longer term - the objective of all competitive strategy development and the common theme of all evolutionary processes.

Advanced Manufacturing Technology

For a definition of advanced manufacturing technology it has not proved possible to remain strictly with the simple DTI view, namely:

'The application of Computers to Manufacturing Operations',

and unequivocably provide for the inclusion of computers in a modelling or forecasting mode. However, even with use of the DTI definition it is clear that a very large range of applications are conceivable for computer systems of varying capacity and that a literature search under this title would assume mammoth proportions.

Further reference to Figure 1 would however suggest that the level of penetration by computers into the sector is relatively small, although it is unlikely that the rate of penetration will have slowed since its publication at the last PRI/Knight Wendling conference.

In seeking an overview of the technology available to the sector it would be inappropriate not to acknowledge the work that has come out of the technological universities, as without it companies short on EDP and manufacturing control system expertise would be severely disadvantaged. The BPF/PED study at Cranfield has provided invaluable assistance to many companies and information from another programme at Bradford that has to be of interest to the contemporary strategist has been used here, (8, 15, 16, 17, 18, 19, 20). More information will probably come from this years' IKV presentation but still more is needed at both the macro and micro level. Much of the output at the microscopic level will rightly find its place in future machine as well as system designs but food for future strategic thought is still contained in the knowledge of its existence and content.

General reviews of AMT, directed towards guidance for manufacturing managers, assume the latter to be seeking to control the production activity - from product design to product distribution, in the most effective manner possible. This will generally produce a list of computer-aided/ computer integrated procedures and for the record these will include:

* Computer Aided Design (CAD)

* Computer Aided Manufacture (CAM)

* Computer Aided Engineering (CAE)

* Computerised Inventory Control

* Computerised Production Planning and Control (normally involving real-time data acquisition from the shop floor and computerised scheduling procedures)

* Computer Numerically Controlled Machining (CNC)

* Direct Numerical Control (DNC)

* Robotics

* Automated storage and issue systems

* Flexible Manufacturing Systems (FMS)

In all the above, concentration is on the generation storage, transmission and integration of data on demand and at high speed, the principal pre-requisite of an optimised competitive manufacturing system.

Although general awareness of the above technologies is acknowledged by most, the proposition remains, that in too few companies within the industry, strategic reasoning is constrained by an incomplete ability to define, as a policy, the form and level of each that should be progressively anticipated by the longer term business plans of sector companies.

In the foregoing the ability to assess the potential impact of the full range of changes capable of affecting the internal and external dynamics of the business operation has not been assumed. It is the potential use of computers in a financial performance oriented forecasting, simulation and modelling mode, as much as their use in attempts to achieve high level adaptive control of the sectors processing procedures that causes the DTI definition to require modification. Whilst the integrated econometric, quantitative process and dynamic factory models for use in the sector are still some way off, the strategic advantages inherent in such systems for the eventual possessors are easy to conceive.

The possibilities visible within an overview of AMT applications might best be seen from a block diagram similar to that shown as Figure 4.

Whilst each manufacturing system design will be unique there is a need for managers to be able to see that such systems of data management are able to reinforce the entrepreneurial flair that will never cease to be required.

It will be seen from Figure 4 that the man/machine interface has not been forgotten and the search for an effective man/machine interactive system has a vital part to play in all competitive manufacturing system development.

If a dynamic control system similar to the high-level feedforward system proposed by Parnaby is configured to cover the 12 month accounting period nominated in Figure 4, strategic objectives as they relate to competitive manufacturing system design will be very ably assisted, once the system is up and running. The high level feedforward system referred to is shown as Figure 5.

For many, the ultimate in terms of advanced manufacturing technology is total computer integrated manufacturing (CIM), which, for convenience, can be considered as a distributed data network having three parts, i.e.

*   CAD/CAM/CAE

*   A Management or Business Information System (which can be linked with the above in a financial modelling procedure, able to integrate business control forecasting and simulation procedures.)

*   A real-time Control system for the business organisation which is order-led and configured using a multiplicity of feedback controls and communication interfaces.

A simplified view of CIM is given in Figure 6 and the strategic benefits have already been well reported(21). At the planning stage it is however necessary to be aware of the fact that as the level of technology involved in any AMT implementation programme increases, the chance of achieving less than the level of anticipated financial return also increases.

Theoretically based heuristic procedures have long been used to advantage in modern production/operations management but significant practical constraints can be expected when control theory is deployed against real time optimisation concepts. It is suggested that in many instances very valuable and workable compromises can be found which avoid frustration yet maintain progress.

For others an engineering emphasis within company culture might allocate priority to the development of a Flexible Manufacturing System (FMS) which can be the most physical and often one of the most visually impressive demonstrations of advanced manufacturing technology. FMS was defined by Ranky(22) as:

'A system dealing with high level distributed data processing and automated material flow using computer controlled machines, assembly cells, industrial robots, inspection machines and so on, together with computer integrated materials handling and storage systems.'

As Ross(23) explained at the first PRI/KW Automation conference, an FMS might therefore range from a cell containing two machines plus a robotics handling device, to a large integrated system comprised of a number of cells, each controlled by its own computer, but controlled overall by a hierarchy of computers. Companies of all sizes are clearly encompassed by such an explanation.

An alternative definition of FMS with a marked mechanical engineering bias was produced by the DTI and for completeness is reproduced below:

> 'Flexible Manufacturing is a system which combines microelectronics and mechanical engineering to bring economies of scale to batch work. A central on-line computer controls the machine tools and other work stations and the transfer of components and tooling. It also provides monitoring and information control. This combination of flexibility and overall control makes possible the manufacture of a wide range of products in small numbers.'

Given the simplified view of a CIM network in Figure 6 and an introduction to cellular structures in the FMS definitions the seemingly popular interest in the progression towards a three tier level of control for work cells of reasonable size i.e. Material, Production and Process Control level, Local Management control level and Overall Management Control level is understandable.

The cellular approach does however stimulate interest in the possible emergence of an advanced processing facility containing a greater assortment of processing methods under one roof than we have been used to.

In assessing the movement of manufacturing technology in the Plastics Processing Sector since the first Conference, we have to conclude that the movement has not been uncomfortably rapid and although information relative to the technology from which development will take place is readily obtainable, it may not be widely held.

However, given knowledge of current technology, or technology that will become available in the short term, system design as it may materialise within the Polymer Processing Sector is not difficult to envisage. As the use of new technology in such systems increases the need for improved human communication also increases. For AMT to be viewed in strategic terms by top management, the extent of its potential coverage and the logic employed in each part of the system needs to be highly visible and equally easily understood. Hopefully it will have been shown here that system logic in graphic form is easily digestable and unnecessary jargon avoidable. Top management involvement in AMT implementation is unavoidable if a project is to succeed. It has however been suggested(24) that whilst a Chief Executive's knowledge of the new technology nominated for use need not be absolute, it is essential that he communicates strategic requirements to a management team selected by himself or the Top Management grouping for their KNOWN understanding of the detail. Such knowledge implies that the Chief Executives of the successful factories of the future will be no technological slouches.

In his paper at the first PRI Conference on Automation Ross stated that the purpose of Automation, and implicitly AMT, was the achievement of greater control over every aspect of production, higher control over quality assurance and improved manufacturing economics. The aim was seen to be long-term

productivity and not short-term payback for the capital investment involved
and the link with competitive strategy is clear. He is not alone in holding
such views regarding investment and in some areas it is felt that to separate the cost dimension from the strategy dimension releases some constraints on progress. Some supporting evidence is reported to exist. A total
decoupling seems difficult to justify however unless the strategic
considerations and implications are severe.

The restoration or improvement in competitiveness achievable from
progression towards manufacturing excellence is clearly significant in
strategic terms if the purposes of automation can be achieved. The fact
that it can be achieved on a progressive basis for many is a bonus.

One year after the last PRI Conference, the Advanced Manufacturing Systems
Group under the Chairmanship of John Osola concluded that investment in AMT
can be self-financing with only a short-term requirement for external
funding. In the polymer processing sector we are not perhaps in a position
to be as unequivocable and any comparison with the 'OSOLA MODEL' would need
to allow for observable differences between the two sectors. Our industry
is however a major one and it is to be hoped that a parallel study with
equivalent financial emphasis can be considered for initiation in the near
term, if suitable control companies can be identified.

Given an accelerating rate of movement on the learning curve and a probable
activity imbalance between companies, strategic reasoning would suggest
that the sector as presently structured will be unable to tolerate the
change. Decision makers will therefore be required to assess the potential
impact of the new technology on their own positions, which, although
defensible now may not remain so in the longer term.

Such is the raison d'etre of competitive strategy formulation in times of
rapid technological change.

Competitive Strategy Formulation

The Classic Approach

Whilst it could never have been the intention or purpose of this paper to
provide a treatise on strategic management, competitive strategy and
competitive advantage, it was thought necessary to identify a framework of
strategic analysis, that could be used in individual approaches to
competitive strategy formulation.

Using three very excellent publications of work on the subject of competitive strategy and competitive advantage by Porter(25),(26),(27) it is
however possible to develop a framework within which the influence of
technology on the dynamics of competition might usefully be assessed. In
view of the use made here of the work of Porter, the model developed is
referred to unreservedly as the 'Porter Model' and the work of Andrews,
Christensen and others (The Policy Group, Harvard Business School) is
already acknowledged in Porter's writings.

Many present may already be familiar with the depiction of the ends (goals) and means (policies) of the 'Wheel of Competitive Strategy' contained in the above references and in the four factors:

* Company strengths and weaknesses
* Industry opportunities and threats
* Broader Societal expectations
* Personal values of the key implementation

which act to limit that which any given competitive strategy can achieve.

Given the extended DTI definition of AMT used in the previous section and some of the examples of computer applications, it can be seen that AMT can be used to influence many of the operating policies directed towards achievement of company objectives, once these have been defined. It can also reduce the strength of some of the limitations indicated above.

We thus have an immediately recognisable link between AMT and part of the classic approach to strategy formulation that has become a standard for many.

In spite of such an acceptable approach to competitive strategy formulation and the publication of a suggested process for formulating competitive strategy, little evidence is found of companies within the polymer processing sector that involve themselves in the depth of analysis that the investigative path described requires. Whilst it is accepted that every firm will have a competitive strategy, the suggestion here is that many such strategies fail to draw benefit from the focus and concentration that is demanded by the planning activity involved in explicit strategy development.

As with all competitive situations, success long-term is associated with knowledge of what an opponent is able to do, rather than what he might do. Such knowledge should cover the longest time period for which analysis is practicable and ideally a tested riposte should be available. From the competitive strategy viewpoint two essential inputs exist, INTELLIGENCE and FORECASTING SKILLS and it is left to each manager here to form their own value judgements in respect of individual company abilities.

The Porter Model - An Introduction

In relating a company to its environment it is necessary to address factors which are applicable to the industry or industries in which it competes and external factors which impinge upon such an industry or industries. Clearly the external factors will affect all competitors within any given industry equally. Whether all can deal with them with equal success is another matter.

The Model suggests that the state of competition within an industry depends on five basic competitive forces whose collective strength determines the ultimate profit potential within the industry - profit potential being measured in terms of the long run return on invested capital.

The profit potential of different industries is expected to vary and the point is made that:

'The intensity of competition in an industry is neither a matter of coincidence or bad luck'

The proposal is that competition in an industry is rooted in its underlying economic structure and goes well beyond the behaviour of current competitors. In addition it is suggested that competition in an industry continually works to drive down the rate of return on invested capital towards the competitive floor rate of return - the return that would be earned by the economists perfectly competitive industry. Most here, who have sought over the years to increase margins will probably accept both proposals from a basis of pained recollection.

In investment appraisal terms, using the association with the risk free rate of return on capital, general acceptance is given to the proposal that should the industry return fall too far below the risk free rate, the incumbent will either be attracted by alternatives or, in the longer term, will run the risk of going out of business. But, as most who have supported rationalisation proposals in the sector know, neither logic nor the basic principles of economics are always seen to apply.

It has already been stated that in relating a company to its environment, it is necessary to address factors which are applicable to the industry in which it competes and it is probably worthwhile to attempt to develop a possible view of the sector from the strategists viewpoint - before continuing with the attempt to link AMT with competitive strategy through the Porter model.

The Polymer Processing Sector

Any researcher attempting to analyse the polymer processing sector soon recognises that the statistical base is weak and this will clearly make life for the strategist difficult. The subject is referred to in a Plastics Processing EDC Report but the report does go on to advise that, some 4,000 companies with active interests in the conversion of plastic materials are likely to exist and that 50% of such companies will probably be involved in on-site processing for in-house consumption. We appear to have less information for the rubber processing sector.

The same EDC report indicates that 80% of the companies in the sector employ less than 100 people, a fact generally confirmed by the Annual Reports of the PPITB.

A debt is clearly owed to the PPITB as the above reports provide a useful guide to the concentrations that exist in the industry and no other document is known to the writer which provides such information in such concise terms.

The strategist will however need to gain more information on the industry through the normal marketing research channels but will not find it easy. This is accepted as being one of the reasons for companies, short on resources and appropriate skills, not performing the detailed analyses that the Porter Model requires. To have a reason for not performing the analysis will not however head off the possible consequences of not undertaking the work.

The PPITB Reports provide an initial insight into the level of diversity in the plastics processing sector and additional secondary data will confirm great diversity in terms of:

* Company size, culture, procedures and objectives
* Processing procedures employed
* Positioning Policy
* Markets and market information
* Market share profiles
* Materials
* Levels of maturity of Company, process, product, markets and materials.

* Parent Company and Ultimate Holding Company relationships.
* etc.

Some companies have however either consciously or unconsciously engaged in extensive segmentation activity and as a result concentrations do occur.

Given the need to establish a better economic understanding of industry divisions a simplified approach to considerations of barriers to entry or exit can be helpful and are easily expressed in matrix form. The simple 2 x 2 matrix shown as Figure 7 can lead to the view, particularly for the trade processor, that history has directed their movement towards the top right hand corner and occupancy of the 'worst case' position. The relatively easy entry in the high growth periods witnessed by the industry has been accompanied by high exit barrier situations for many companies during recession. The inevitable result is excess capacity, with an accompanying general decline in profitability as prices are forced down by people having nowhere else to go.

Directional policy for a company might be assisted by reference to Shell's 3 x 3 matrix where support can be found for:

* Niche position strategy
* Phased withdrawal from an industry sector by holding companies
* The need for double or quit responses.

The use of matrices will obviously reflect individual interpretations of the inputs and outputs and, although the polymer processing sector is a difficult industry to analyse, matrix application might mean that analysis need not be stopped at a point which tends to serve only short-term interest.

Wilkinson(28) has used matrices and the PORTER MODEL to investigate the interface between Corporate Strategy and R & D Planning and the direction of this work could be put to good use within most companies within the sector, when competitive advantage is being sought.

It is important to realise however that the point being made is not that
the use of matrices provides possible answers but that strategic analyses
can be conducted to provide preliminary guidance in industries where the
database at first sight appears limited or fragmented.

Given that investment requirements may constrain the smaller company, the
entry barriers to a new style of industry by implication may have been
raised. If existing barriers to exit are lowered by the economic harshness
of the products of economies of scale, the reaction to investment proposed
by ACARD and ROSS is indirectly given substance in strategic terms.
Following any industry shake-out and high volume growth and investment a
position appropriate to the bottom right hand corner might become
perceivable for the sector.

The Porter Model of the Forces Driving Competition and AMT - A Linkage

A recap on the purpose of automation, as expressed by Ross at the PRI
Conference in 1984 re-introduces the perceived need to achieve:

*   Greater control over every aspect of production
*   Higher control over quality assurance
*   Improved manufacturing economics

To this we could add perceived needs for increased differentiation, which
might in turn involve improved supply flexibility, improved productivity
levels associated with new product introduction etc. The present AMT
capability is therefore well correlated with the above stated needs and
purpose.

Given the block diagram which depicts the five forces driving industry
competition, Figure 8, it can be seen that a framework exists against which
any particular element of AMT can be tested - in terms of its potential
impact on competitive strategy.

Industry Competitors

Whilst there are strategic implications in the manner of operation of all
four of the principal functional areas of an enterprise, the strategy/AMT
link with which we are concerned here causes a primary emphasis to be
placed on operations management and the use of the productive system as a
competitive weapon. We thus return to the price-cost gap, the use of
quality and market orientation as the stage for AMT in the search for
competitive advantage.

A strategy must be developed that links the available producing technologies with the needs of the market in a way which ensures long term defensibility for the chosen position of the enterprise in the market place. It
is therefore relevant although some license is being taken, to refer to the
experience curve phenomenon where evidence confirms that as experience is
gained through production, unit costs are reduced.

Once only attributed to a learning effect by operators, the curve is now recognised as having additional inputs which include:

* Changes in production methods (including layouts)
* Product design (for manufacturing purposes)
* Economies of scale
* Improved flexibility
* Improved levels of standardisation.

Previously the experience curve has been associated with product focussed systems and is formalised mathematically as:

where:

$$C_n = C_1 \, n^{-b}$$

C_n = the cost of the n'th item

C_1 = the cost of the first item

n = the cumulative output in units

b = a parameter depending on the rate of unit cost decrease.

Where license is being taken is that within the polymer processing sector experience gained with a wide variety of products in process focussed systems can be aggregated.

If this can be accepted, another facet of the experience curve phenomenon can be considered to apply. This states that if through process technology advantages, a firm can establish itself on a lower percent experience curve compared to a competitor, it will have lower unit costs even if both firms have the same cumulative output.

The use of the aggregation approach to the experience curve tends to reduce its greatest limitation, product obsolescence. Process technology advantages can therefore be seen as a viable route to competitive advantage, providing such process advantage represents a true advantage in financial terms. This rider tends to account for the re-stated reluctance to totally decouple the cost dimension from AMT studies, except where the strategic considerations are severe.

It is therefore suggested that we have a view of a link between AMT and competitive strategy through the product and process advantage the former can create.

Potential Entrants

Any link between AMT and competitive strategy in this area might need to consider:

* future growth prospects in the sector

* acquisition possibilities within selected market areas

*   the possibility of structural change within the sector and possibilities for influencing such change

*   the level of feasibility associated with an advanced processing enterprise.

Whilst access to financial and human resources is clearly required, the range of technology available for discussion at Automation '86/K86 etc. and the product and process advantage it can engender in a growth sector, could well make assessments of feasibility attractive for potential new entrants seeking corporate growth, diversification or both. The fact that an advanced polymer processing unit can be approached on a progressive basis is likely to enhance the likelihood of its eventual appearance. Clearly, the established companies require to be aware of such possibilities and it is unlikely that AMT could be excluded from any defensive strategy.

Pressure from Substitute Products

Within the polymer processing sector substitution is seen to be more of an opportunity than a threat although no place for complacency exists. In any proactive response, however, well developed and inter-related activity between marketing, R & D, and product design will seek to use the total AMT resource available to it to secure competitive advantage.

Vertical integration and resort to in-house manufacture by a previous consumer can be considered a substitute product by the trade processing sector and the response required is similar to that which can be applied to reduce buyer power. Again, the total AMT resource available to an enterprise will require to be summoned to its defence.

An in-house processor can also be put under pressure by the availability of products from an advanced supply side industry able to provide advantage to his own competitors. He must therefore remain aware of all such changes and the new technology that might be demanded.

Bargaining Powers of Buyers

For some buyers, changes to manufacturing system design within their own operations has made 'in-house' manufacture more meaningful. For those that remain in the market for product, any forecast of increased competitive forces on the supply side increase buyer power. Aggressive responses by the supply side should therefore attempt to promote decisions favourable to themselves by invoking the pre-stated purposes of AMT and influencing the price mechanism, quality and supply flexibility.

The tapered integration policies of some major end users of the products of the polymer processing sector, where extensive use of AMT is often visible at on-site manufacturing locations, clearly require processing technologies to be matched if views on competitive prices are to coincide.

Bargaining Power of Suppliers

The tactical approach to limiting supplier power lies in the creation of alternatives, particularly in situations where an enforced sole supplier status is conceivable.

For many processors emphasis will need to be placed on the need to create alternatives for the supply of materials and labour, with AMT being clearly linked to both. Whilst dependence on labour might be reduced in volume terms it needs to be recognised as a resource requiring a quality input and training policies will need to be found adequate.

Generic Competitive Strategies

Competitive strategy development using the PORTER MODEL as a framework requires responses to be available to cope with the five forces driving competition in an industry.

The PORTER MODEL suggests that three generic strategic approaches are available which are identified in the references provided as:

*   Overall Cost Leadership
*   Differentiation
*   Focus

All three are directed towards outperforming competition in an industry and whilst a mixed strategy is considered feasible, it is necessary to be fully aware of the factors responsible for the adoption of such a policy.

With a brief indication of the relevance of AMT to the forces driving competition already given it is now necessary to assess its relevance to the generic strategies to achieve the objective contained in our title.

Overall Cost Leadership

Low relative cost is the predominant theme, with the cost reductions available from the experience curve being vigorously pursued.

Low relative cost is seen to be an adequate defence against all of the forces driving competition and an ability to defend such a strategy is stated to require:

*   Optimum productive performance

*   High relative market share

*   The ability to design for manufacturing

*   A well structured product range

*   Heavy and continuous capital expenditure in state of the art
        equipment.

The low cost position lays claim to evidence suggesting that the high
financial margins required can be achieved and that market share does not
have to be bought.

To pursue such a strategy is however to risk, amongst other things:

*   Over-extension of all resources

*   Scrapping assets still able to provide an 'acceptable' return.

*   Market opportunities missed through the emphasis placed on cost.

*   Cost inflation that produces vulnerability in a specific product
        in a specific market.

Differentiation

An industry wide acceptance of uniqueness in product or service is that
which is sought and although costs cannot be ignored they are of secondary
importance in strategic terms.

Again, evidence is available which suggests that such a strategy offers a
viable defence against the forces driving competition and although high
returns are seen to be possible, market share can suffer.

It may commonly need high quality design facilities, detailed product or
service knowledge and a high level of customer orientation.

To follow such a strategy reportedly risks amongst other things:

*   Lack of attention to cost
*   Product or process imitation

Focus

Focus as a strategy needs no definition but can embrace concentration
placed on targets extracted from a wide range of alternatives i.e. product
markets, geographical markets etc.

Whilst differentiation in a broad sense is inevitably achieved this can be
through pursuit of one or both of the preceding strategies in specific
instances.

In pursuing such a strategy an enterprise risks either being out-focussed
on one hand or the narrowing of differentiation on the other.

The Generic Strategies and AMT - A Linkage

In choosing to focus more upon competing through manufacturing one is drawn towards an assessment of the link between AMT and the low cost strategy. In this area the pre-stated purpose of AMT implementation is more highly visible and a link through the experience curve sustainable.

No doubt the following papers will also confirm such a link.

The impact of AMT in the product design area with its ability to aid both low cost and differentiation strategies will also be demonstrated during the conference and it is hoped that in casting a personal perception of the scope of AMT against an elementary strategic framework intuitive acceptance has been replaced with a more tangible rationale.

Investment Appraisal

With a paper on financial control systems in the non steady-state situations included in the programme the above subject will not be extensively debated here.

The preference of Ross and the ACARD report for a longer term view on returns has already been referred to and other views which relate to the decoupling of the cost dimension from the decision-making process will have been noted.

This paper has also made the point that it is doubted whether sole consideration of the risk free rate of return will be found to be adequate and an acceptable logic is sought which has due regard for the level of perceived risk.

From the strategists viewpoint it can be suggested that a company, that does not anticipate becoming exposed to a high level of risk over the longer term, can retain a strongly positive link with the risk free rate during investment appraisal.

A company faced with a threat from a competitor happy to accept a lower rate of return in the short-term, in exchange for long term security will, all things being broadly equal, be immediately placed under pressure to reduce emphasis on the risk free rate.

Both companies, given that they are UK companies, would be placed in the same dilemma should a foreign competitor seek to destabilise the market for long term gain.

Once again, one has to be encouraged by the Osola Committee findings but financial justification might, in the short-term, become highly correlated with re-appraisal of the extent to which the uptake of AMT can ensure that one lives to fight another day.

Hopefully, the process of re-appraisal will be aided by this, the previous and following papers.

REFERENCES

1.   WALLECK, S. - Restoring Our Manufacturing Entrepreneurship

2.   FOYER, P. - Integrated Scheduling and Handling Systems. Coventry Consortium Seminar, November 1985.

3.

4.   KANTROW, A.M. - The Strategy Technology Connection. Harvard Business Review, July-August 1980.

5.   SCHUMPETER, J.A. - The Theory of Economic Development. Harvard University Press, 1934.

6.   SIMMONDS, T. - New Technology in the Plastics Processing Industry. Plastics Processing Industry Training Board, 1984.

7.   HAMBLIN, D.J. - Manufacturing Management and Control Systems in The Polymer Processing Industry. SERC/PED, British Plastics Federation, 1982.

8.   COATES, P.D. - Micro-processors in Polymer Processing. Shell Polymers, Volume 8, Number 3, 1984.

9.   ADVISORY COUNCIL FOR APPLIED RESEARCH AND DEVELOPMENT (ACARD) - New Opportunities in Manufacturing. The Management of Technology, 1983.

10.   PLASTICS PROCESSING EDC - Plastics Prcessing, The Next Ten Years, 1983.

11.   PLASTICS INDUSTRY TRAINING BOARD - Annual Reports for 1983 and 1984.

12.   DEMPSEY, P.A. - New Corporate Perspectives in FMS. Ingersoll Engineers, 1983.

13.   OSOLA, V.J. - Advanced Manufacturing Technology, The Impact of New Technology on Engineering Batch Production. NEDO Advanced Manufacturing Systems Group, 1985.

14.   POWELL, A. - FMS and CIM In Japan. Production Engineer, January 1986.

15.   KOCHAR, A.K. and PARNABY, J. - The Choice of Computer Systems for Real Time Production Control. Production Engineer, October 1977.

16.   HAMMOND, A. and KOCHAR, A.K. - Role of Computers in Production Planning and Control for Plastics Products Manufacturing. Plastics and Rubber Processing, June 1978.

17.   PARNABY, J., BATTYE, P.G., HASSAN, G.A. - Computer COntrolled Injection Moulding and Extrusion. Plastics and Rubber Processing, September 1978.

18.   HOLLWEY, M.W.M., KOCHAR, A.K., PARNABY, J. - Using a Real-time Mini-Computer for Production Control. Production Engineer, November 1978.

19. PARNABY, J. - Concept of a Manufacturing System. Int. J. Prod. Res., Volume 17, Number 2, 1979.

20. HASSAN, G.A. and PARNABY, J. - Model Reference Optimal Steady State Adaptive Computer Control of Plastics Extrusion Process. Polymer Engineering and Science, April 1981.

21. DESCHAMPS, J.P., GUNN, T., KRAUS, I., MAIER-ROTHE, C., MURPHY, A., STOKES, P., QUINTRELL, J. - The Strategic Benefits of Computer Integrated Manufacturing. Arthur D. Little Inc., 1983.

22. RANKY, P.G. - The Design and Operation of FMS. IFS Publications Limited, 1983.

23. ROSS, D.S. - Advanced Manufacturing Technology - Evolution or Quantum Leap? PROC. PRI/KW Automate or Liquidate Conference, Brighton, 1984.

24. WALLECK, S. - White Glove Technology in Blue Collar Plants.

25. PORTER, M.E. - Competitive Strategy. The Free Press, 1980.

26. PORTER, M.E. - Competitive Advantage. The Free Press, 1985.

27. PORTER, M.E. - The Technological Dimension of Competitive Strategy. Harvard Graduate School of Business Administration Working Paper, 1983.

28. WILKINSON, A. - Corporate Strategy as a Source of Ideas. R & D Research Unit, Manchester Business School.

FIGURE 1

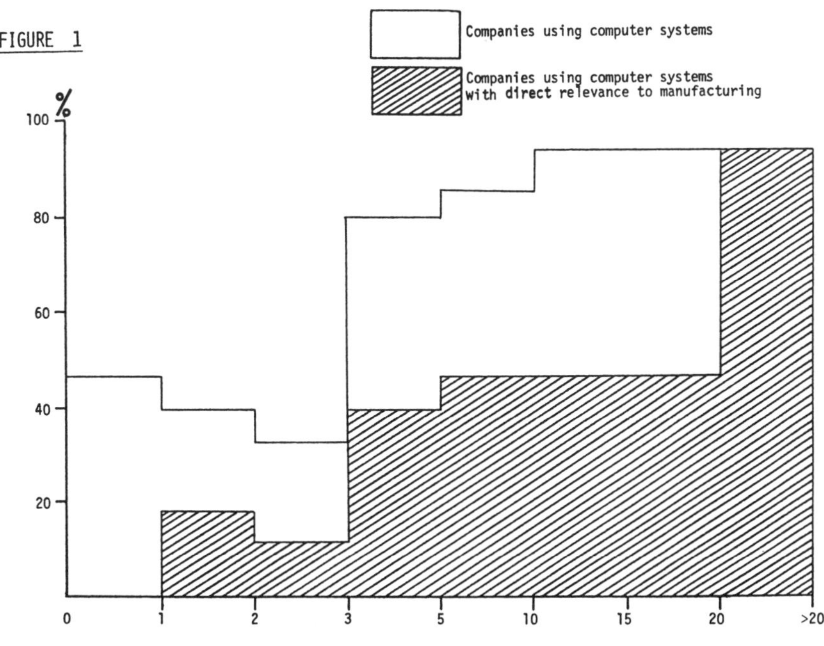

COMPUTER USAGE BY POLYMER PROCESSORS
SOURCE: "MANUFACTURING MANAGEMENT AND CONTROL SYSTEMS IN THE UK POLYMER PROCESSING
INDUSTRY" (SERC/PED - BPF PUBLICATION NOV. '82)

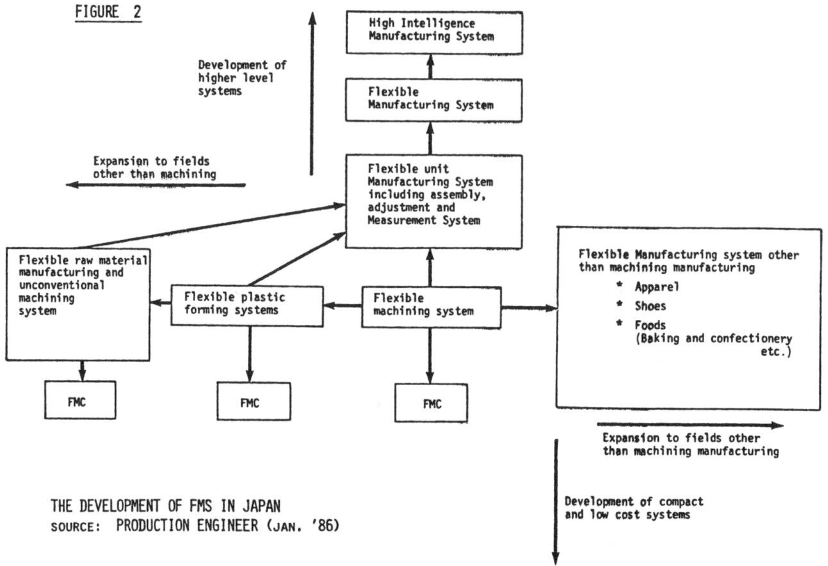

THE DEVELOPMENT OF FMS IN JAPAN
SOURCE: PRODUCTION ENGINEER (JAN. '86)

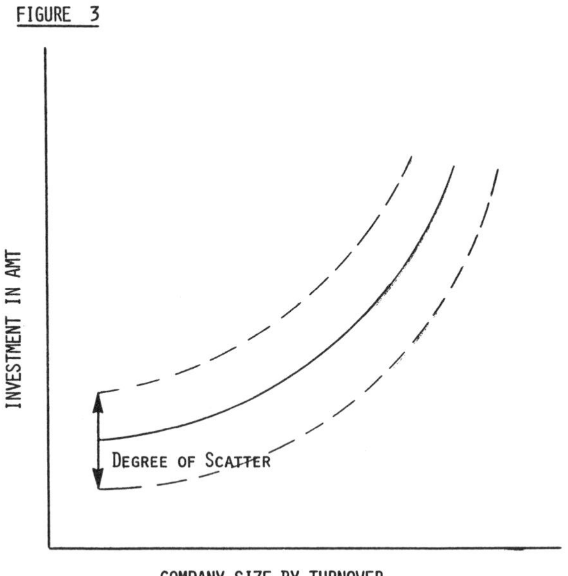

FIGURE 4

INITIAL 'UNFETTERED' VIEW OF GOALS
FORECASTS AND PREDICTIONS OF MARKETS
DESIGN PARAMETERS OF PRODUCTS AND SERVICES
FORECASTS OF:
- TRENDS, SEASONALITY AND IDEAL MIX OF PRODUCTS AND SERVICES
LONG RANGE BUSINESS PLAN   TERM - 3 to 5 years
* Competitive Strategy Development
* Technology related issues
* Secondary projection of "realistic" goals
* Rough-cut Resource Planning and Projected Returns
MANUFACTURING SYSTEM DESIGN
TECHNICAL REQUIREMENTS OF PROCESS
HUMAN RESOURCE REQUIREMENTS
MANUFACTURING PLAN   TERM - 1 year
* Tertiary projection of "essential" goals
* Competitive strategy execution
INPUTS
PROCESSING
OUTPUTS
FEEDBACK
P R E D I C T I O N
RESULTS FROM:- Casual forecasting methods
- Time series forecasting models
- Predictive Methods
- Craft type programs
S I M U L A T I O N
RESULTS FROM:- Management and control systems development
- Database generation
PROCESS PLANNING & CONTROL
RESULTS FROM: -
- On-line, real time
materials management system

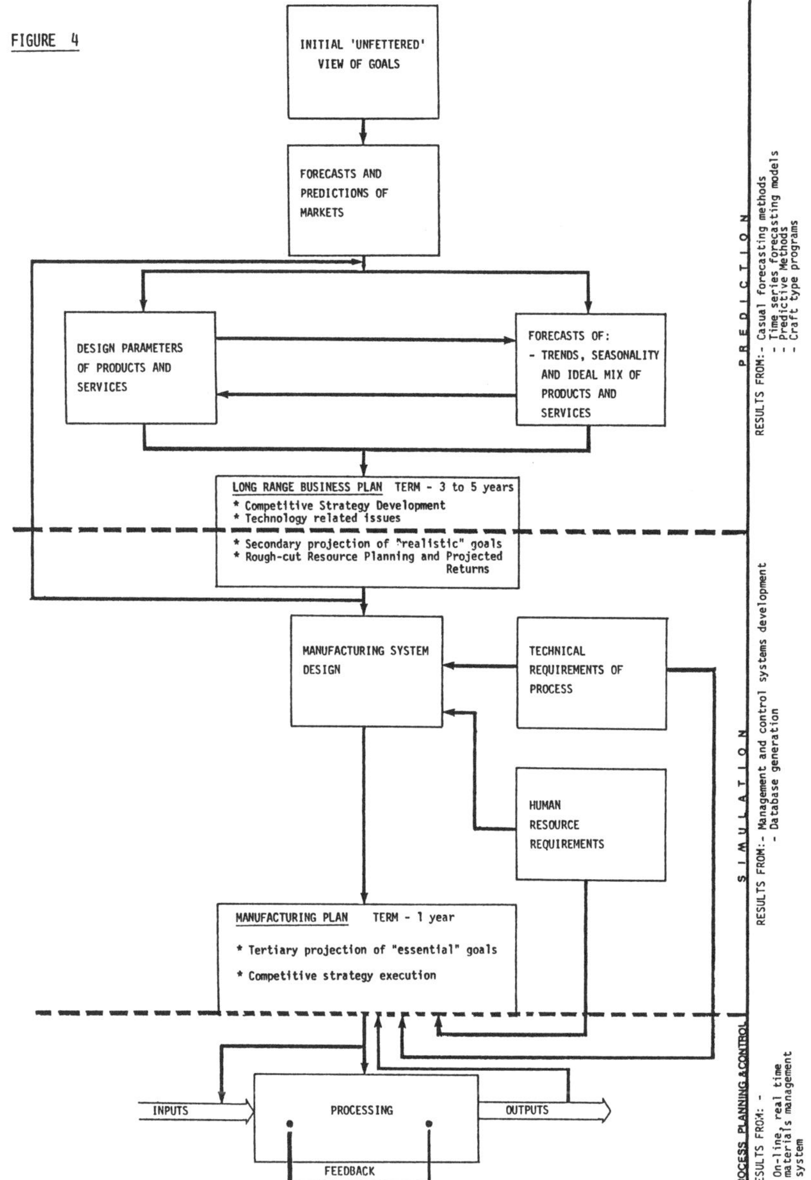

FIGURE 5

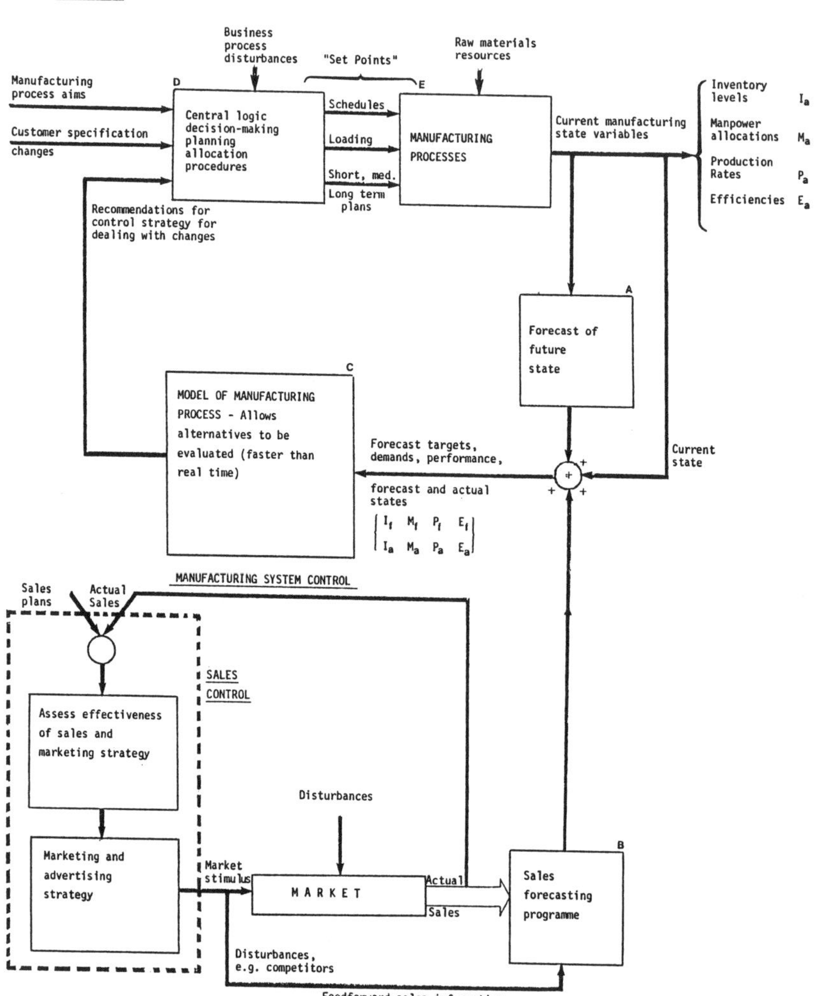

HIGH LEVEL FEEDFORWARD CONTROL BLOCK DIAGRAM

SOURCE: "CONCEPT OF A MANUFACTURING SYSTEM" - J. PARNABY, 1979

FIGURE 6

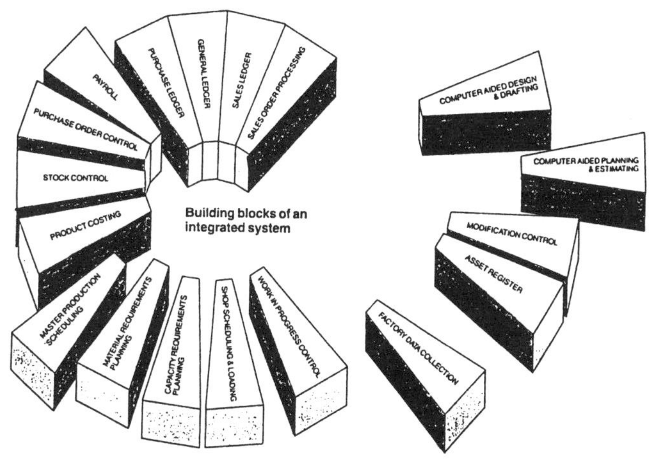

SIMPLIFIED VIEW OF CIM

FIGURE 7

EXIT BARRIERS

	LOW	HIGH
LOW	Low Stable Returns	Low, Risky Returns
HIGH	High, Stable Returns	High, Risky Returns

ENTRY BARRIERS

SOURCE: M.E. PORTER. "COMPETITIVE STRATEGY" 1980

FIGURE 8

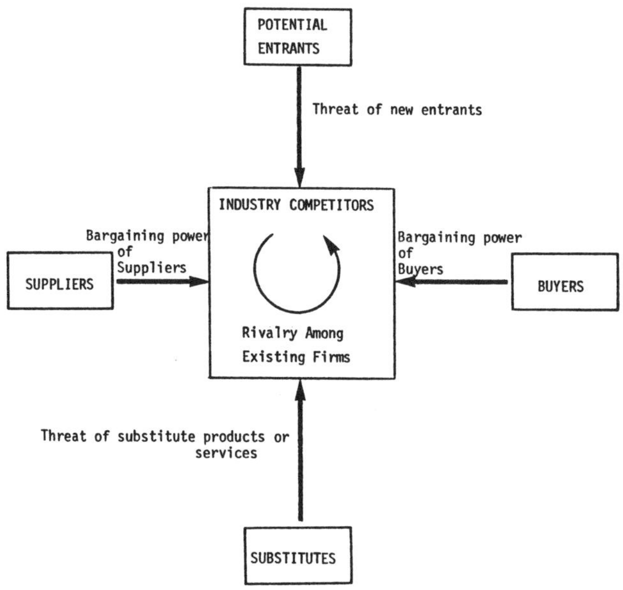

FORCES DRIVING INDUSTRY COMPETITION
SOURCE: M.E. PORTER "COMPETITIVE STRATEGY" 1980

A STRATEGY FOR THE DESIGN OF MANUFACTURING SYSTEMS WHICH MEET BUSINESS NEEDS

Brian Small*

The plastics industry has been experiencing a period of rapid growth, but manufacturing systems have often developed piecemeal. A strategy is needed which will satisfy business needs for competitiveness, quality, responsiveness and cost. One approach is to apply 'Just-in-Time' production techniques for maximum efficiency, and also use advanced manufacturing technology where appropriate. Introduced over five years in planned stages, a combination of these methods will enable a company to balance investment with cash flow and ensure that changes will be absorbed with minimum disruption and maximum benefit.

INTRODUCTION

Rapid growth has been a feature of the plastics industry in recent years. An increasing variety of products has become available, the range of applications is continually widening, and demand is rising. However, this apparently healthy market is also highly competitive, and few companies are making substantial profits. Ironically, the very speed of expansion in the industry has contributed to a lack of profitability: manufacturers have eagerly grasped at every new business opportunity, and a high proportion of companies are now attempting to supply to a wide sector of the market. As a result, high variety, low volume production is often the norm, and production methods, business systems and investment policy have developed piecemeal, in response to market demands as they have arisen. In short, manufacturing strategy does not meet business needs, and there is a imbalance between costs and the prices the market will bear.

There is thus a need for managers to step back and consider a change in basic strategy. To increase profitability, it will be necessary to aim more precisely at particular markets and products - in other words, to become more specialist. This involves choices, investment, and management of change. In this paper we examine the implications of this strategy for the design of manufacturing systems in the plastics industry if they are to meet business needs in the future.

BUSINESS NEEDS

There are four principal needs which manufacturing strategy must satisfy if a company is to be successful and profitable: products must be competitive, quality must be high, delivery and repair must be prompt, and costs must be kept to a minimum. We now examine each in turn.

*Managing Director   Ingersoll Engineers Inc

Products Must Be Competitive

Somewhere in the world is your most effective manufacturing competitor. Perhaps there are several in the first division. Imported goods flood the market, and even newly industrialised countries are producing some goods which are equal or superior in quality to those made in the UK, and lower in price. Understanding what your competitors achieve is the essential prerequisite for any plans for renewal or improvement.

To do this, you need to examine the problem using a simple three-stage approach. The first stage is called positioning. It entails asking yourself all the questions needed to understand perfectly where you are in the world market for your product. This does not mean just knowing everything about your, business and his, and the approach he takes to it compared to yours. Its much more than you normally expect to do in a market survey, but it lets you set targets which can be the objectives of further work. These targets can relate to what is enough to beat the competition, what is possible, and what is achievable.

The second stage is concerned with inward critical analysis of your company and its activity - and here an outside consultant, preferably with market knowledge, can provide valuable objectivity. To do it alone can be difficult.

The third stage is then to determine the actions necessary to place yourself in a dominant position in the marketplace.

It is no use companies seeing themselves as institutes - mighty and unchanging. Business managers must broaden their horizons and be aware of all the pressures which influence strategic and tactical planning. You must send your manufacturing people out into the marketplace to assess the most appropriate indices for measurement of competitiveness. These may be:

- Sales per employee
- Instant availability of spares
- The degree of specialisation
- The number of potential customers who are changing to a policy of buying instead of making

The two last points merit further discussion. First, in considering the appropriate degree of specialisation, it is necessary to manage the product range. Careful selection of the target market will allow the processor to match product structure with the processes he chooses to undertake. For a product owner, this means achieving customer requirements for end-product variety from a minimum library of components and tooling. This may include provision for combining components as far as possible in order to improve performance and reduce cost. For a processor, making other peoples' products, it implies skill in die design, in order to rationalise tool components and maximise flexibility between machine types. It also implies careful planning of material specification and sourcing. At present, users often say they would like more advice from manufacturers both in retail design and in making the best of the tooling options available. Manufacturers who select areas for specialisation can find their advice becoming more valuable and sought-after, and this is a distinct competitive advantage.

In developing the 'make v. buy' possibilities for business, a producer or processor can become more competitive by tailoring his product more closely to the customer's specific needs. One way to do this is to look for opportunities to increase the value of products - for example, by making complete sets of parts as packages, or assembling them. Not only does this increase profit margins, but in addition, the manufacturer provides an attractive service which customers may not find readily available elsewhere.

Only when the manufacturer has analysed his competitiveness carefully in this way will his company have a firm international competitive base.

The Need for High Quality

Customers expect to be able to use plastics without any further finishing being necessary, but this standard of finish is not always achieved by current manufacturing methods. There is a high risk of losing business if quality control is inadequate, particularly in such areas as manufacture of PVC, composites and functional components, but responsibility for quality control in most factories rests not with the workers responsible, but with a separate department.

A worker who produces a large batch of components may neither know nor care that 10% of them are defective. Another worker whose job is to fit two components together may simply scrap any defective ones because he can draw from an ample supply. Quality control people are responsible for finding out what causes defects and for deciding what parts should be reworked, and so workers have no reason to be interested in solving quality problems. In addition, some manufacturers seem to believe that provided an 'acceptable' level of quality has been achieved, there is no need to devote resources to raising standards further.

In the last few years, a totally different concept of quality control has been developed. To use new technology and 'Just-in-Time' production (JIT), manufacturers must aim for <u>total</u> quality control. This idea is based upon five main principles:

-    Reject the ideal of 'acceptable quality levels'. Aim for <u>continual</u> quality improvement.

-    Concentrate on component quality necessary to achieve very high first-time quality finished/assembled products.

-    Make each person responsible for the quality of his or her own work.

-    Use measures of quality that are simple and visible.

-    Develop suitable automatic measurements of quality.

The benefits of this approach include increased worker motivation and higher productivity levels as well as substantial improvements in product quality.

Prompt Response and Delivery

Failure to respond rapidly to orders and long lead times for deliveries are problems not usually caused by lengthy production processes. In fact, the most time-consuming aspects of manufacturing are the delays between each stage of manufacture, or in the set-up time for the machines. Set-up time is the most disruptive interruption, as it forces the creation of inventories all over the factory, or in the store, so that assembly or the customers can be assured of a relatively steady flow of products or components. Another source of delay is lack of balance. For example, if one machine operates at 20 parts an hour and the next at 10 parts an hour, there is a built-in interruption ahead of the slower operation, and parts wait until equilibrium is restored.

Other delays are caused by quality checking, machine breakdowns, and waiting for parts delivery, tools and fixtures. Even plant layout can delay production, for if consecutive processes are located some distance apart and if space has to be provided for storage, then the opportunity for interruptions, the possibility of damage and the tendency to accumulate work-in-progress all increase. Layout problems may relate to an entire plant as well as to individual operations within a plant. Certainly in most factories, it is accepted that parts are transported significant distances between stores and operations. In a typical factory, progressive additions will have been made as the years have gone by and the result is that, after a few years, workflows become extremely complicated.

All these interruptions (often described as "unavoidable delays") may hold up work to such an extent that processes requiring a few minutes' work in three or four stages may take weeks to pass through the factory. Delivery becomes uncompetitive, and the cost of carrying inventories soars. A company may allow for labour delays of as much as 15% - 20% when calculating the cost of products, and some operatives even receive incentive payments for "waiting" during interruptions.

To improve response times and deliveries, it is often advantageous to use a JIT approach. Each process can be planned to lead on smoothly to the next, and although it may not be possible to eliminate all interruptions, it is certainly possible to minimise them. An ideal JIT system is configured so that the time required for parts to flow from goods receiving through the shop and to despatch is only a little longer than the time it takes to complete all the value-adding steps in the manufacturing process. Application of modern process and tooling technology may be valuable, particularly in relation to improving changeover times, and speeding the flow of information.

Scheduling can be made more efficient by instituting systems which can improve liaison with both suppliers and customers. It is important to encourage the growth of partnership, so that, ideally, suppliers respond rapidly by pre-arranged agreement. Customers then receive a better product or component and an improved delivery service which encourages then to give firmer guarantees of future business, or to place orders for products with more added value. To help everything to fit together - orders, scheduling, supplies, manufacturing and delivery - improved information systems are needed, and computers may well have a role once the company has identified precisely what it needs to integrate.

All this will help the company to manufacture its products more smoothly and quickly so that it becomes possible to schedule small batch production economically. Flexibility is thus greater, and the company's attractiveness to customers is increased.

<u>Keeping Costs to a Minimum</u>

In the plastics industry, the cost of materials is a higher percentage of total costs than it is for, say, metal manufacturers. If raw materials account for 50% - 75%, or even more, of the selling price, product and die design need to make the most economical use of them. It is also vital to reduce wastage to a minimum, especially as recycled materials in significant quantities can reduce the quality of the product. Total quality control is therefore highly desirable in terms of both cost reduction and customer satisfaction. The use of advanced manufacturing technology to this end becomes an economically viable proposition if it helps prevent the creation of much expensive scrap, and prevents sub-standard products from going out to the customer.

The other main thrust of the drive to reduce costs is the attack on wasted space and excess stocks. Storage of materials, be they scrap, raw materials, or components awaiting assembly, is costly in terms of overheads. If factory space is too valuable to be taken up with waste materials, it is also too valuable to be taken up by non-productive uses such as storage of high levels of inventory. This is particularly important in the case of high value raw material, which often ties up capital unnecessarily.

By adopting a JIT approach, companies can release a great deal of capital for more productive use, or reduce interest charges on borrowing. Instead of keeping stocks of everything 'Just-in-Case' it is required at short notice, the company's policy changes to making only what it can deliver at once, buying only what it can use at once, and finishing products immediately after moulding, so that materials remain in the factory for the minimum time. The information systems referred to earlier are vital here, and a well organised JIT approach can reduce inventories to near zero, as well as minimising investment in such non-process equipment as silos, mixing facilities and warehousing.

These, then are the main business needs which manufacturers in the plastics industry need to take into account. We now consider some practical aspects of planning and putting the chosen strategy into effect.

<u>CONSTRUCTING A FIVE - YEAR PLAN</u>

Because every business is different, a tailor-made plan is necessary to maximise business gains by applying the most appropriate technologies available. Advanced manufacturing technology (AMT), computers and integrated manufacturing systems may well be useful tools, but they are not the starting point. A plan which puts technological advance first will normally be a business disappointment. Technology is a means to an end, but first the end must be clearly defined.

The first requirement is to prepare a five-year plan relating the needs of the business to the technological possibilities, and to set up models showing the economic and technical benefits of particular courses of action.

In considering the future development of a company, it is important to ensure that the plans made fulfil the needs of the evolving business and avoid imposing a grandiose design which might disrupt the effectiveness of existing production. A five-year plan should take this need into account while providing the direction and structure within which small incremental developments can take place. This will help to prevent the error and inconsistency associated with piecemeal investment.

Having evolved a design for the overall plan, a company's management then needs to establish business milestones for the five-year period. Four or five of these need to be set, probably each with two or three goals to be achieved. The next task is to establish the areas of action and the improvements needed to achieve each milestone. The models will normally indicate a number of possible approaches requiring evaluation, each including hardware purchases, new control systems and software, procedural improvements, and the development of infrastructures such as layout, organisation and communications.

It is at this stage that the use of techniques such as integrated manufacture and advanced manufacturing technology can be considered in detail. Having established the aims for the next five years, management can assess how cost effective it will be to automate particular processes and how feasible it will be to combine then into a continuous manufacturing system.

Most of the discussion so far has been about best manufacturing practice and, indeed, radical thinking about the problems of controlling businesses can achieve a great deal without recourse to AMT. Significant reductions in overheads, materials, indirect labour, floorspace, work-in-progress and inventories are all valuable gains to be made in this way.

However, the evidence from recently implemented modernisation programmes shows that the merging of the skills of managers and technologists provides the best solution to the problems of survival and future progress. This is especially true for strategic planning, because it enables the company to balance investment with the benefits to be derived and the company's cash flow. Over the five years, then, the aim is for a <u>combination</u> of best manufacturing practice and advanced manufacturing technology to make the company more successful, profitable and firmly-based in its market.

<u>USING INTEGRATED MANUFACTURE</u>

Fully integrated manufacture relates to the entire manufacturing business; it is not just about production. When faced with increasing competition, a company cannot simply concentrate upon improving specific aspects of the factory processes. The methods of running the business - from order intake to despatch - must be the very best that can be envisaged. This requires meticulous planning, and if a business needs to improve radically in terms of cost, delivery, quality and competitiveness, then the planning logic must start from fundamentals, the three important steps being first, simplify; secondly, integrate; and thirdly, apply modern technology as appropriate.

The need to <u>simplify</u> relates to a number of areas of manufacture which in most companies are over-complex:

-        Product design
-        Manufacturing processes
-        Handling of products from goods inwards to final despatch
-        Layout of factories and production lines
-        Organisational and administrative activities
-        Information flow

The advent of automation provides a unique opportunity to <u>integrate</u> these aspects of manufacturing into a continuous process. However, to gain the maximum return on investment, it is important not to treat automation as merely machine replacement. If this happens, the specialist function or department concerned is treated in isolation from the rest of the manufacturing process, and all the existing complexity, delay and resulting overhead costs remain.

In approaching the task of integration, it is vital to identify interfaces carefully and to ensure the transitions between them are effected as smoothly and as rapidly as possible. The 'Just-in-Time' principle, so effectively applied in Japan, is usually considered in relation to the movement of work materials only. But it can be applied also to the movement and storage of tooling, fixtures and other materials, to the work of the purchasing or despatch departments and, most importantly, to the generation and supply of information required in manufacturing. There are many ways of achieving integration by non-technological means: these include departmental consolidation, plant consolidation, product range management, and more flexible working practices.

However, in many areas, the <u>application of modern technology</u> is likely to be helpful and cost-effective. Computer-integrated manufacture is a possibility, as is investment in rapid changeover equipment. The business information systems may well need updating, and existing computer-based systems will possibly be quite inappropriate. The approach that suits one department or function may be irrelevant to the needs of the company as a whole.

Do not overlook the importance of <u>people</u> to your manufacturing system; at every stage in its planning, implementation and operation, success or failure will depend upon the skill with which you manage the people concerned. Training will be needed to provide the skills needed, as over the five-year period of transition, there will probably be a major switch from a majority of largely untrained people to a preponderance of highly trained staff, in addition to a fairly rapid reduction in manning levels (in the absence of rapid sales volume growth). The sequence and timing of implementation must be planned to make this transition work progressively. This helps to reduce aggravation as well as risk. The people aspect of major projects is the most neglected at present, and yet it takes the most planning, judgement and skill to manage well.

The main objective throughout should be to complete everything in the simplest and most timely way, and to ensure that each stage fits into the overall strategy as a building block contributing to the final structure.

CASE STUDIES

Having considered the broad principles of strategic planning, we now look at two particular examples. It can be both interesting and instructive to consider how companies have, in practice, approached their individual planning problems in designing manufacturing systems.

1.   A manufacturer of electro-mechanical assemblies was successful in selling an attractively styled, low-priced instrument pack to a major customer. However, the company had to find a way of increasing production rapidly and substantially - from a maximum of 6,000 packs a week to 10,000 new style packs a week within four months.

The priorities in planning were to achieve management control of output and quality, to manage work-in-progress and all inventories properly, and to allow assembly operators to work more effectively. The major decisions taken were to:

- Integrate as much work as possible vertically under one supervisor.

- Use parallel working rather than series (operator work content of 5 minutes per unit instead of less than 30 seconds).

- Provide a line production controller and make him responsible for all material, from goods receiving to finished packs and for liaising with materials provisioning.

- Concentrate work in as small an area as possible to give good geographical control of people and parts.

- Use kitting rather than trackside parts feeding to ease changeover, reduce clutter and improve quality by providing better control of parts. Three kitters were made responsible for the control of parts.

- Place rectification in an area near the supervisor to give him control over defects and the issue of substitute parts.

- Improve product testing; in the medium term move towards computer control, and introduce an audit process.

- Introduce a major vendor management programme over 1 - 2 years, to improve price, quality and delivery.

- Emphasise ergonomics and the general working environment.

The project constituted the first part of an overall review of the facilities at the site, and radical improvements were achieved within the four months:

AREA OF BENEFIT	BEFORE	AFTER
Throughput time (complete pack)	30 days	5 days
Inventory turns	4	9
Delayed deliveries	20%	0%
Inventory	£1.2 million	£0.55 million
Direct minutes per unit	6	5
Rework	20%	<5%

The cost totalled £250,000 - including capital investment - and the benefit was confirmation of a contract worth £7 million (25% of site sales). The new facility had a major effect on customer confidence in the product, and management was able to go on to set even more ambitious targets for the future over a longer time-scale.

2. A second example demonstrates the need for flexibility in planning implementation and shows how careful design of improvements to meet business needs can ensure that they are not only cost-effective, but even self-financing.

An oilfield equipment manufacturer wanted to plan and implement a low-cost and fast-throughput structural assembly facility. It was decided to close existing buildings and move into a new, expanded facility to support the growing business. The facility plan developed challenged traditional approaches to issues such as equipment and processes, layout and workflow, product design, and organisation. After implementation had begun, the market suddenly began to decline drastically, so that equipment orders had to be cancelled, and the move to the new site had to be completed using a mixture of old and new equipment. Then later, when market conditions were favourable, the plan was up-dated and a team established to resolve major problems.

These problems crossed many departmental lines:

The new facility was not considered a continuous line production (raw material to despatch), resulting in:

- Solving the wrong problems
- Problems not being easily visible
- Imbalance in the lines
- Parts being <u>pushed</u> to despatch rather than <u>pulled</u>
- Foreman not being assigned to a line

An informal scheduling system was used by the shops, resulting in:

- Delayed shipment
- Unknown capacity
- Late parts
- East yard buffer inventory
- Handwritten paperwork (causing lost time and poor productivity)
- A 'hot list' for scheduling

Reports were not used to identify and help solve the recurring problems, resulting in:

- Ad hoc solutions
- Paper being generated but not used

Utilisation of raw material and Just-in-Time manufacture for each line were not adequately balanced, resulting in:

- High lot sizes to use material
- Additional material handling
- Additional inventories
- Higher costs
- Split orders/loss of control

Workers were not flexible or cross-trained within the line concept, resulting in:

- Imbalanced lines
- Additional inventories
- Late parts

No measuring tools were used such as:

- Manufacturing cost
- Cost per unit
- Number of fork lift trucks
- Truck delays and reasons for truck delays
- Assembly hours per unit
- Structural hours per unit
- Total components not available to ship one day before shipment date
- Detail components not available to ship one day before shipment date
- Inventory turns
- Orders received summary
- Analysis of changes
- Comparison of actual versus planned

The plans to deal with these problems were still based on a mixture of old and new equipment, but the following changes were implemented:

- Modification to allow continuous line production (raw material to despatch)
- Simplified scheduling system to improve delivery and minimise inventories
- Improved painting technology
- Improved material handling and storage, minimising the use of fork lift trucks
- Improved forecasting
- Improved management information

The annual value of the improvements was $10 - $15M a year, and the cost for achieving this was $9M. However, in balance sheet terms, the cost was minus $2M. The one-time inventory reduction of $11M more than paid for the entire project. The company's bottom line was improved by $10M to $15M a year, and they got it for nothing!

KEY GUIDELINES

Some key guidelines to developing manufacturing systems which meet business needs are:

- Develop a precise understanding of the business and the market to which the manufacturing facility and associated systems have to respond - and its position in the total supply chain. Assess the likely future changes.

- Achieve a balance of advanced manufacturing technology and best manufacturing practice

- Establish a carefully considered conceptual manufacturing approach for the whole factory/plant/site/business. Establish a suitable facility and systems outline plan which is not constrained by current activities (usually planning five years ahead).

- Determine the steps necessary to reach the ultimate goal and decide how many steps can be taken at a time.

- Acknowledge that the plan will change, find ways of overcoming constraints, and avoid becoming boxed in. If the plan is so inflexible that it cannot be changed, what chance is there of achieving a truly responsive manufacturing environment?

- Settle for nothing less, in the long term, than an integrated system. Build on a solid data base foundation, aim for a well structured modular approach and seek objective advice, but remember that the responsibility for installing the system and making it run cannot be abdicated.

- Be prepared to invest substantial time effort and resource (people and money) in tailoring the system to specific requirements. This applies to the physical plant and equipment as well as to the computer/control systems and software.

- Devote as much time to planning the implementation as to product design and development.

- Remember that people are the key. Encourage an attitude of common sense, hard work and co-operation that puts overall business and operational requirements ahead of section/personal interests.

ADVANCED MANUFACTURING STUDIES AT IKV AACHEN AND SOME EUROPEAN EXPERIENCE TO DATE

G. Menges and G. Bolder*

Modern, up to date production of plastic components
must atune itself to the prevailing economical requirements. The reduction of stockage leading to so-called
just-in-time production, small orders requiring the
possibility to change the production with short setting
times and the increasing demands on the product quality
are important aspects here. A satisfying solution for
these problems must include consistent application of
automation means, up to date measurement and control
technology as well as handling systems or robots. In this
report it is intended to discuss solutions for complete
systems developed by the IKV, using examples of injection moulding, SMC-processing and extrusion.

INTRODUCTION

In the development of plastics processing machines the future
clearly belongs to fully automatic process control and increased flexibility of the production. Witheach of the two
aspects the processing technique has to be given further consideration.

The experience and conceptional ideas of the IKV in the
matter of flexible production will be discussed by means of
three examples originating in the fields of injection moulding, SMC-moulding and extrusion. The solutions are based on
present knowledge in machine technology, measurement and
control engineering and previous concepts of automation.

Today, however, one is aiming for systems which include preparation, refinishing and assembling all the way to the finished product and also utilizing all possibilities of fully
automized operation. Presently, almost every single manufacturer of plastics processing machinery works on the solution
of these problems. Thus the production centres realized by the
IKV stimulate the industrial application and give new impulses
to the large-scale realization.

FLEXIBLE MANUFACTURING CENTRES FOR INJECTION MOULDING

Automation has made the most obvious industrial progress in
the machine technology of injection moulding. This progress
has been based upon new developments in the field of machinery as well as on the application of the latest computer
technology and the analysis of the injection moulding process
itself.

* IKV Aachen

The modular structure of injection moulding machines is a particular advantage here. Thus the machine can be adapted to the specific requirements of the respective moulding or material. In practical manufacturing the flexibility resulting from changing of moulds and plastification units is complemented effectively by handling systems for removal of the mouldings and by metering devices. A process computer is responsible for the coordination here. Up-to-date versions of such computers are not only used for control of the operation sequence, but also perform all control functions concerning injection speed, pressure and temperature. Presently, a number of companies are realizing the possibilities and advantages of holding pressure-control by means of the pvt-model developed at the IKV /1-3/.

In addition to what has already been said, it is important to production planning and quality maintenance to integrate the single machines into an overall process control system. Almost all manufacturers of machines are working on this problem today. They are also confronted with the difficulty of standardizing these networks here /4/.

The moulds as well have been subject to further development with respect to flexibility in the last few years. For thermoplastic and elastomer materials the development of hot runner and cold runner injection moulding has therefore been carried on with respect to automatic removal of mouldings.

However, the full flexibility is only realized if the production can be changed over to new products within a very short time. Many manufacturers of machinery have taken up the idea of quickly changeable moulds or cavities and have come to different solutions here /5, 6/. The main difficulties lie in the handling -particularly in the case of large moulds - and in connecting the power supply.

The following tasks were defined for the flexible injection moulding centre developed at the IKV:

- fully automatic continuous operation of the plant, even for mould change of a unit
- reduction of tooling times
- manufacture of a thermoplastic/elastomer composite moulding with less assembly steps
- straightforward cleaning of the elastomer mould
- complete assembly up to the finished part.

A plant was designed particularly along the lines of these tasks; it consists of the following components:

- thermoplastics injection moulding machine
- elastomer injection moulding machine
  (both of them with a mould changing system)
- handling robot
- assembly stand
- process computer

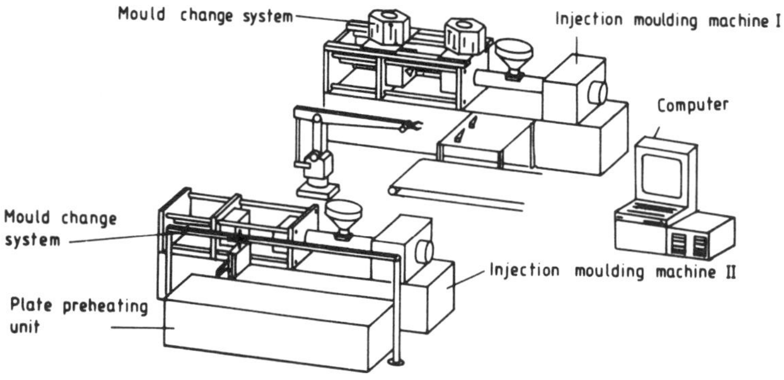

Figure 1: Arrangement of the components of the manufacturing
         centre

An arrangement such as the one illustrated in figure 1 is used
for production of composite parts, for instance, ice scrapers
with rubber lips or hooks with rubber suckers /7/.
The thermoplastic inserts are moulded in injection moulding
machine I. The robot removes the parts and delivers them to
the elastomer machine which moulds the elastomer element to
the thermoplastic part. At the end of the cycle of machine II
the robot removes the finished part and puts it into a magazine for packing; the magazine may then be transported by
means of, for example, a band conveyer.

Both machines are equipped with a mould changing system. Thus
it becomes possible to produce composite parts - as it is described above - as well as pure elastomer or pure thermoplastic
parts. In this configuration it is indispencible to do the latter, because the cycle times of the two machines can not always
be synchronized in such a way as to realize continuous production of a composite part in several machines. Generally the
cycle times will differ. This brings about the necessity for
short-term storage between the machines in order to make allow-
ance for these differing machine cycles. In particular this is
an essential necessity for the changing of mould plates required for product changes or for cleaning contaminated moulds
of the elastomer machine. Otherwise the entire manufacturing
system would be put out of operation, while, instead, it was
possible to continue the production with at least one machine.

Certain assumptions were made in terms of size and type of
mouldings for the design of the mould changing system in order
to eliminate the expensive, complicated change of entire moulds.

Many different moulding geometries can be classed under single
so-called moulding families an account of geometric similari-
ties or approximately equal filling volumes. Within such moulding families it is not necessary to change the entire mould.
It was therefore undertaken to develop a mould concept which
allows the mould plates in the mould to be changed. The corresponding mould consists of a bolster into which the different
mould plates can be slided-in. The temperature control system
as well as the runner system (and the ejection system respectively) are an integrated part of the bolster. A revolving
magazine such as the one shown in figure 2 is used for storage
and reception of the exchangeable mould inserts. The magazine
is attached to the machine by means of a supporting frame. Rotating the magazine and moving the mould inserts into operating
position are performed pneumatically. Heating and cooling of
the insert plates is provided in the operating position as well
as in the stand-by position. This ensures that the production can
be continued without any start-up scrap after exchange of an
insert plate.

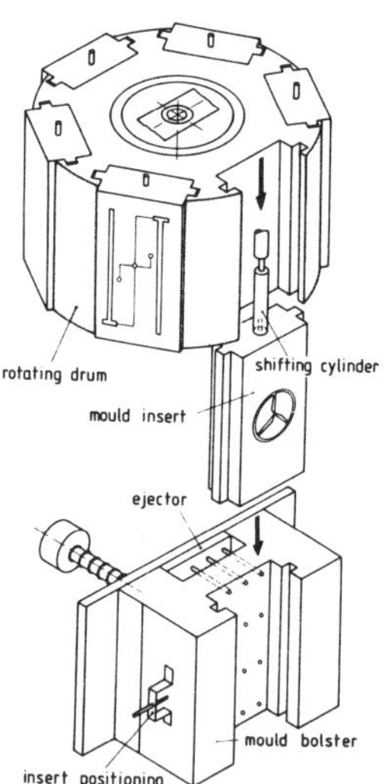

Figure 2: Mould changing system

Mould changing systems, such as the one already existing in thermoplastic injection moulding, are of even greater significance to elastomer processing, because they are even more promising in terms of rationalization here. So far commercial systems are based upon the idea of changing the entire mould unit. In the case of thermoplastic injection moulds this may be a tenable conception since they do not have to be cleaned constantly. Rubber moulds, however, have to be taken completely apart for cleaning. All band and cartridge heaters, including the leads as well as all sensors and all other delicate mould components which do not withstand the cleaning baths, have to be removed. It is therefore essential to design rubber injection moulds in such a manner as to minimize or entirely eliminate the disassembly necessary for cleaning the mould.

As a consequence, an optimum conception, from the engineering point of view, must be based on the idea to exchange, by means of a changing device, only the forming mould components instead of the entire mould. So far the design of most rubber injection moulds already satisfies the requirements of such a conception to some extend, because it also aims at a most straightforward mould structure.

The basic design of the mould coincides to a large degree with conventional injection moulds; commercial standards can be applied for the construction of the bolster. Heating of the mould cavity plates is performed indirectly by means of two heating plates. The heating power of these plates is determined in such a way as to be able to despense with additional heating elements in the forming plates. With this configuration the often expensive ejection equipment is not necessary either, because external robot or handling system are applied for demoulding the parts. If the handling system is not able to apply the necessary demoulding forces by itself, it is additionally possible to use compression air to initiate the separation of the moulding from the mould wall. Pressure sensors, thermocouples and load cells as well have been re-located in areas which are not part of the mould plate. As a consequence there are really only two "naked" steel plates between the heating plates; they are attached to the mould bolster by means of a cocking cylinder. In the case of contamination or a necessary product change, these steel plates can be removed laterally from the mould within a short time.

The mould plates are then pushed into a magazine arranged at the side of the machine (figure 3) form where they can be delivered to a cleaning station, for example. The magazine serves to pre-heat the mould plates when they are not used in the production cycle. It can be displaced parallel to the axis of the machine.

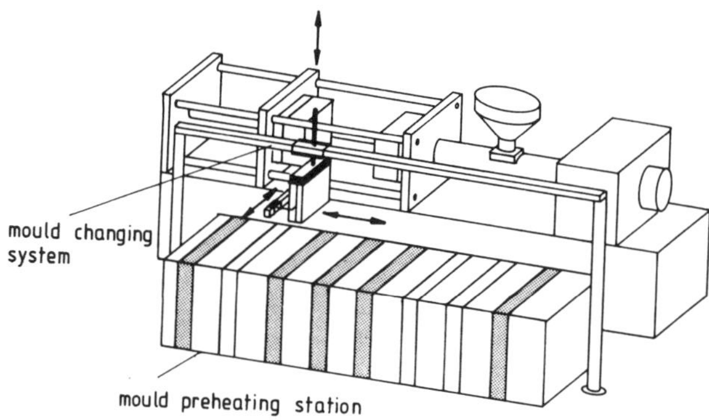

Figure 3: Injection moulding machine with mould changing system and preheating station

In some cases of product change the manifold has to be ex-
changed in addition to changing the plates. To that end the
split level of the mould can be relocated between the clamping
plate on the side of the nozzle and the supporting strips by
means of an automatic rechucking device; this brings about that
the cold runner is exposed and can be changed in the same way
as the mould plates. Such a concept is suitable for moulding
families consisting of parts of approximately equal dimensions
and which, in this special case, do not feature high undercut.

Robots that can be programmed by means of the TEACH-IN method
gain more importance in the field of flexible manufacturing
systems. They can be installed, for instance, on the side of
the injection moulding machine. For flexible manufacturing
such robots are an economical solution, if they are in the
position to perform a number of tasks. Such tasks could be the
removal of mouldings from different machines as well as the
subsequent performance of several equations such as, for example, the assembly of mouldings. Further appropriate tasks
may be:

- spraying of the mould with release agent
- inserting parts into the mould
- removal of the sprue
- selective discharge
- refinishing
- cleaning of the cavity
- assembly of several components
- packing.

It can be assumed that all operation can be carried out by one
and the same gripper. It is therefore important to include the
possibility of automatic gripper changing systems into the robot concept. A distributed system performs the control of the
entire manufacturing centre. In it, only computer systems communicate with one another, such as master computer, machine
control computer, mould changing computer and robot computer.
Functions for the control of the machine, the optimization of
the holding pressure phase on the basis of the pvt-model and
the acquisition of operation data are a constituent part of
this computer system as well.

EXPERIENCES WITH FLEXIBLE PRODUCTION IN THE FIELD OF
SMC-MOULDING

During the past few years evidence has been obtained in a
wide variety of different ways on the suitability of the
SMC-moulding process for large-scale production. In Europe,
for example, fenders and hoods have been manufactured in large
numbers for the car industry /9/. The pioneers here are the
American car makers, who produce an even wider variety of car
body parts of SMC and seem to want to extend the number of
applications appreciably /10/. One point of discussion that
always arises, however, is the lack of automation in the process. Although solutions have been suggested from various sides
as regards full automation /11/, as far as we know, no-one has
yet put these concepts into practice. The primary reason for
this is almost certainly the enormous cost of such a plant.

A concept for a highly automated production system was therefore developed from the many years of work done by the Institute in this field. The main features of this concept are its
relative simplicity and its modular structure. The modular
structure makes it possible, on the one hand, to build up the
system over a longer period of time, and on the other hand, to
integrate part solutions into the existing production sequence.

Subdivision into the following aspects is suggested:

- preparation of charges
- loading of the press
- moulding
- part removal
- refinishing.

The work on a flexible manufacturing centre for production of
SMC-components began with purchase of a modern short stroke
press made by the firm Müller-Weingarten /12/.

Figure 4 shows the basic construction of the press. Its particular feature is that the rapid closing of the press is
separated from the actual compression stroke. This is achieved
by having the closing unit moved by separate hydraulic cylinders. During the downwards movement, the closing unit is halted
briefly before making contact with the material and is fixed by
clamping devices by friction to the guide columns.

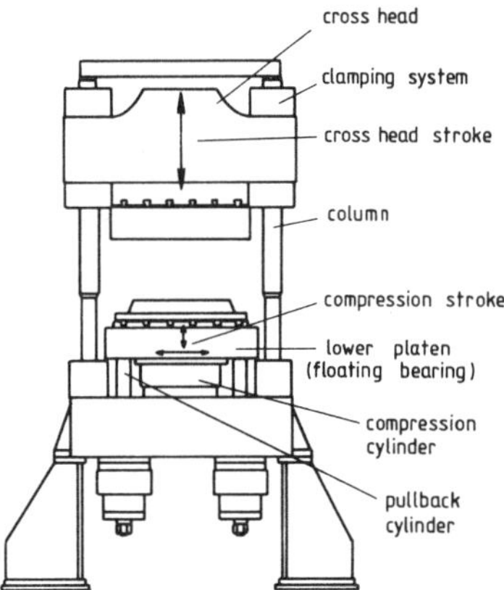

Figure 4: Short stroke press with parallelism control

The compression stroke is now carried out via the compression
table from below, with the latter being operated by a very
short-stroke hydraulic cylinder with a low oil volume. This
brings considerable advantages as regards the control technique in view of the low compressibility of the oil column
in the short-stroke cylinder. Pull-back cylinders are located
on each of the four corners of the table and they permit the
parallelism of the stroke to be controlled to about $\pm$ 0.025 mm,
even in the case of eccentric load application.

We have developed a very simple handling system to automatically load and unload the press. It works on the principle that a carriage, which moves along linear rails and
is provided with grippers to grab the final part and the
charge, enters the open press. Figure 5 shows the principle
of the system.

The linear rails 2 are positioned to the side of the mould
and are fixed to the guide columns of the press via clamping elements. The height can thus be adjusted, so that
they can be adapted to different mould heights. One linear
rail is replaced by a pneumatic cylinder 4 (without piston
rod), which looks after the drive-in and drive-out operations. Similar cylinders are installed for the vertical

movement of the inner part of the carriage 7, so that with
the aid of vacuum suckers, the part 11, which has been lif-
ted up with the upper half of the mould, can be gripped.

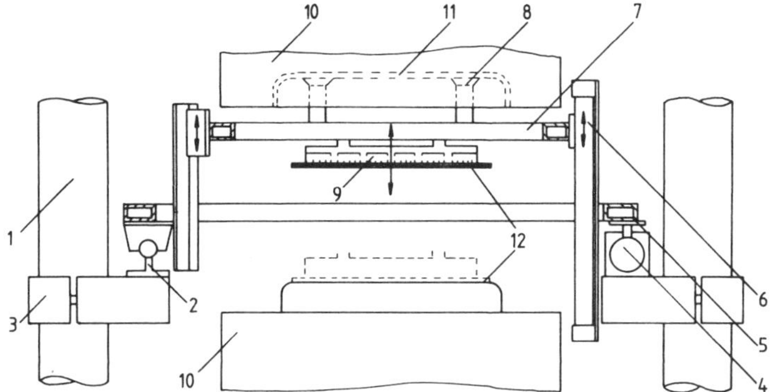

Figure 5: Handling device for loading and unloading of an
          SMC press.

The charge on the ejectors is then put down. For this, use is
also made of a vacuum-operated gripper 9, which uses honey-
combs as the contact surfaces because of the stickiness of
the charges. The charges are taken from a magazine; the
sealing film is removed with a specially developed device.
After moulding it will be unavoidable to carry out some refinishing or secondary treatment such as the removel of
flashes, drilling of holes or milling of cut-outs. According
to our experience this task can be appropriately performed
by a robot which is also able to carry out the assembly of
the finished part /13/.

A suitcase is manufactured as an exemplary product. Its two
halves are made of SMC. Thermoplastic injection moulded
mountings complete the suitcase.

All motions of the press are controlled by servo valves in
closed control loops. They ensure that the press performs
reproducibly and accurately in the way given by the freely
progammable compression velocity and curing pressure profiles. The charge-over point from compression velocity
control to curing pressure control depends on the compression
force here. This technique allows for significant reduction
of flash formation.

Last not least it should be mentioned that the press process
control - a unit developed by the IKV - also offers the
possibility of acquisition and control of process data.

<u>FIRST CONCEPTS OF FLEXIBLE PROFILE EXTRUSION</u>

In the production of extruded semi-finished products the
pressure of the situation also forces the processors to
produce smaller lot sizes. Otherwise the multitude of manu-

factured profiles would entail the necessity for extensive
stockage. Moreover it would not be possible to supply
customers with urgent orders fast enough during the times
when a large-volume production is being conducted.

Flexible production means drastic changes to the extrusion
process particularly if one considers that response times are
very long and that it is hence mostly required to operate the
process stationarily. The customary, long setting times and
the start-up problems are a great set-back to the economical
advantage of such a production, unless the technical concept
is particularly matched to these new requirements. The
start-up of the extrusion process can be accelerated significantly by the means of up-to-date measurement and control
engineering /15/.

Furthermore supervision of the manufacturing process makes
it possible to improve the production with respect to
dimesional accuracy - in other words the product quality.
The tooling time can also be reduced by changes concerning
the machinery itself. A production line for profiles constitutes a suitable example for an adapted, flexible solution. Figure 6 shows the set-up of such a plant.

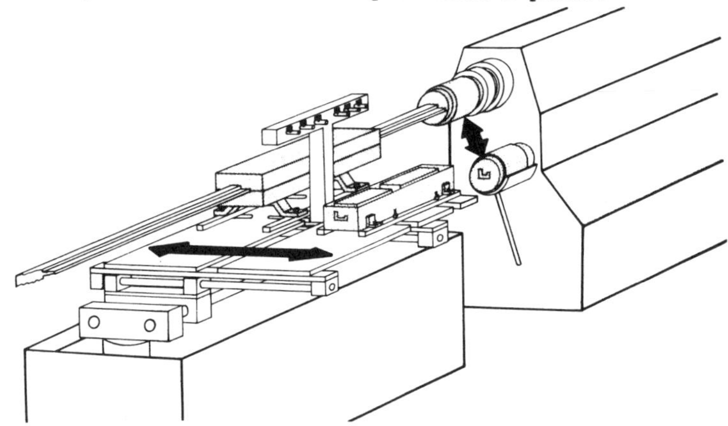

Figure 6: Semi-automatic profile and calibrating die changer
          unit

Various profiles are manufactured with an up-to-date PVC twin
screw extruder, changeable dies and sizing devices. Changing
of the die is accomplished manually; it is possible to
preheat the die. While changing the process computer positions
the sizing unit in the line. Thus the extruder can be started
again immediately after the die change and the new profile
can be taken off.

The computer is able to predetermine all working points of
extruder, take-off and temperature control for the chang-
ing-, the tooling-, the start-up- and the production-phase.
With these simple methods the interruption of the production

can be reduced to a few minutes. Of course the machine setter
is busy with cleaning the old die, preparing a new die and
sizing unit for the next change, while the production is
already running.

Measuring and control techniques are an essential part of
fully-automatic profile extrusion. Modern sensors such as the
ones used in non contact measurement of the profile width by
means of a CCD-camera allow information on the process condition to be acquired directly on-line. The knowledge of the
correlations in the process brings about the possibility of
controlled interference /16/. Figure 7 illustrates the set-
up for the control unit.

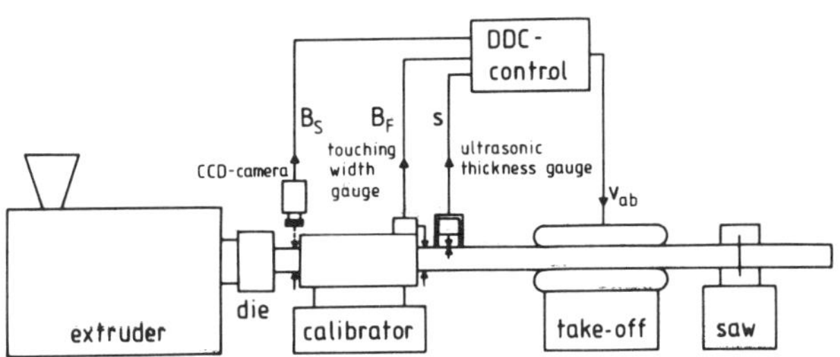

Figure 7: Dimension control at the profile extrusion line

This control deals with two major problems, the selection of
sensor positioning and the control of a delay system.
Measurement of significant geometric dimensions ahead of the
take-off makes it possible to compare actual and desired
values; such dimensions are, for example, the width of the
profile measured by means of inductive touching width gauge
or the profile thickness measured with an ultrasonic probe.
The severe disadvantage of this measurement, however, lies in
the long delay time between the measuring devices and the die
as the place where the material is still plastic and can
hence be influenced by the take-off. This leads to the
problem of difficult delay time control.

A preliminary measurement behind the die eliminate this
disadvantage. Thus relative measurement of the profile width
can add to a fast control if the final dimensions are used
for subsequent corrections. The structure of the control is
represented in figure 8. This cascade control system features
an internal control loops which reacts very fast to changes
of the profile width behind the die by adjusting the take-
off. The controller can be designed either as P- or as
PI-controller; due to the short response time there are no
dimensioning problems here. The master controller on the
other hand has to be adjusted to the delay-time of the

measurement section die – measuring device. The delay time
changes with the take-off speed – hence also with the working
point of the plant. According to our investigations a
flexible scanning time of $T_{s1} = f(v_{ab})$ has proven to be
appropriate. With this it becomes possible to give the
process computer one set of parameters evaluated by an
optimization /15/.

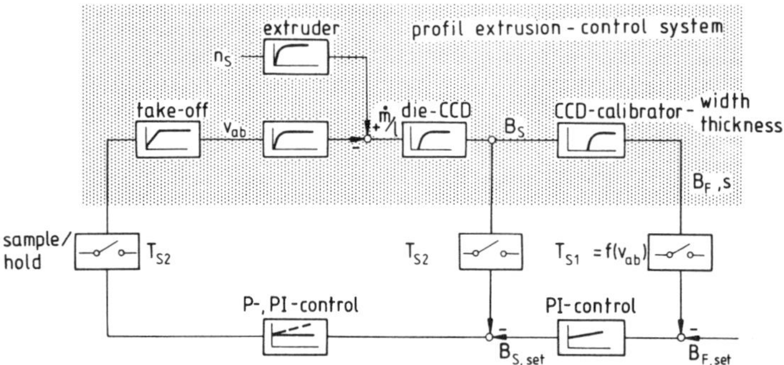

Figure 8: Cascade control of a profile extrusion-line

Figure 6 represents the behaviour of the control in case of a
discharge rate disturbance of 20%. With the extruder running
with starved feeding, the feeding rate was altered suddenly
here. The reaction of the take-off rate as the correcting
variable is very obvious. Neither the auxiliary controlled-
condition B_s nor the controlled condition s show any sign of
overshoot. The wall thickness s is controlled with an accuracy of ± 0,03mm even if considerable discharge fluctuations
occur. The main dimensions were mentioned constant with an
accuracy of less than 0,1mm for all investigated profiles.
In practical operation the discharge fluctuations are half as
high, at the most, if the hopper is maintained full.

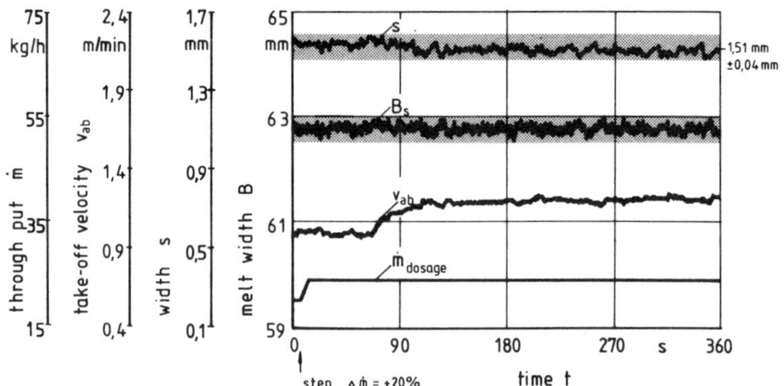

Figure 9: Disturbance behaviour of the control

The controlled hopper metering may be an alternative to the
described control as it is offered for example by Battenfeld
/17/. However figure 9 shows that this type of feeding may
entail slower transmittance. The ultrasonic thickness measurement which is offered by this firm in pipe extrusion corresponds
to a concept comparable with the CCD-camera and can be
operated with similar control engineering elements /17/.

With the concepts for flexible automation introduced so far
it becomes evident that a number of interesting results have
been obtained already and that flexible manufacturing will
become more and more important in the future.

REFERENCES

/ 1/ Menges, G.   Injection Moulding - moves towards process control
     Matzke, A.   Plastics and Rubber Institute
     Janke, W.   Conference Preprints
                      8/1-8/15, 1985

/ 2/   pvT-process control
                      information from Sandretto

/ 3/   Das pvT-Konzept für konstante Formteilqualität durch
                      Nachdruckoptimierung
                      Firmenschriften der Fa. Battenfeld

/ 4/   PV-Umfrage bei Spritzgießmaschinenherstellern
                      Plastverarbeiter (36) 1985, Nr. 10, S. 122

/ 5/ Jäger, F.-M.   Maschine und Umfeld bei automatischem Wechsel der
                      Spritzgießproduktion
                      Seminar: Flexible und automatisierte Fertigung und
                      Montage von Spritzgießmaschinen, Aachen, 1984

/ 6/ Langecker, G.   Das Battenfeld-Werkzeugschnellwechselsystem Mouldfix
                      Beitrag zum Battenfeld-Symposium mit Sonderschau
                      vom 4. - 7.5.1982

/ 7/ Weyer, G. u.a.   Herstellung von Gummiformteilen - wie könnte eine
                      moderne Fertigung aussehen?
                      13. Kunststofftechnisches Kolloquium 1986, Block 10

/ 8/ Matzke, A.   Prozeßrechnereinsatz beim Spritzgießen - ein Beitrag
                      zur Erhöhung der Flexibilität in der Fertigung
                      Dissertation an der RWTH Aachen, 1985

/ 9/ Lenk, D.   Funktionsteile aus SMC im Nutzfahrzeugbau am Beispiel
                      von Stoßdämpfern und Schallkapseln
                      19. Jahrestagung des AVK, Freudenstadt 1984

/10/ Wood, A.S.   Compression moulding fine-tuning the technology
                      Modern Plastics International, March 1985

/11/ Büchle, G.   Die Massenfertigung von SMC-Teilen
                      Kunststoffberater 5/1984

/12/ Schmelzer, E.   Formteile mit Langfaserverstärkung - Produktion
     u.a.   13. Kunststofftechnisches Kolloquium, Aachen 1986, Block 2

/13/ Neise, E.   Verarbeitung von faserverstärkten Kunststoffen mit
                      Industrierobotern
                      Dissertation an der RWTH Aachen, 1986

/14/ Cherek, H.   Dissertation an der RWTH Aachen, 1986

/15/ Bolder, G.   Ein Beitrag zur Massedurchsatz- und Temperaturführung
                      des Extruders
                      Dissertation an der RWTH Aachen, 1984

/16/ Breil, J.   Automatisierungsmöglichkeit an Rohr- und Profilextrusionsanlagen
                      Dissertation an der RWTH Aachen, 1986

/17/ Fischer, P.   Neue Wege für die Automatisierung der Rohrextrusion
                      Firmenschrift der Firma Battenfeld

FINANCIAL CONTROL SYSTEMS FOR PRODUCT COST EVALUATION AND MANUFACTURING PROJECT JUSTIFICATION

Clive A. Turner*

An examination of the financial implications of engineering decisions in respect of capital expenditure procedures and product costing control systems. The main theme is to emphasise **UNDERSTANDING** of financial implications and to examine the consequences of process choice, product portfolio and business organisation expressed in terms of financial risk and revenue expenditure.

INTRODUCTION

It is recognised that extensive work has been published examining the weaknesses of Financial Control Systems; the short-comings of engineers and absence of Manufacturing Strategies. The paper includes reference to this work, however, the main emphasis of the paper is to examine the business team's accountability for the evaluation of investment decisions and the determination of subsequent product costs.

The whole subject area is addressed from the perspective of a qualified accountant who, for a number of years, has worked in management consultancy and who accepts the lack of development of accounting techniques and understanding over the last 60 years. Therefore, the main emphasis of this paper is to debate a new framework for:

i) the understanding of revenue expenditure and its subsequent control processes

ii) the role of the engineer in manufacturing based decisions

iii) the role of the accountant in manufacturing based engineering decisions

iv) business team accountability

* Turner Associates

The need to underpin the understanding of manufacturing based solutions
with the statement of the financial implications is paramount to the
business view of the consequent business risks and related management
controls.

Product costs are the consequence of the process choice effected in the
investment decision; the product related expenditure (eg raw materials;
bought-out parts and/or processes etc) and the structure based
expenditure. Identification of product cost must be supported by
appropriate cost control systems.

Management accountability for capital and revenue expenditure should be
localised and supported by post-audit procedures.

<u>There is no quick-fix solution.</u> Every effort should be made to avoid the
easy route of attributing blame
"Its the accountant imposing impossible targets"
"Its the Marketing Department - the promised orders never materialised"
"Its the engineer who purchased too much excess capacity".

Instead the business team accountability for determining strategy and
underpinning such strategy with strategy-based investment decisions
leading to the understanding of revenue expenditure and appropriate
financial control systems.

COMPANY A

Finance Director: "We have operated this method of accounting for
more than 20 years, we have agreed the method with our main customer:
we are profitable .... we don't believe we need to change".

Factory Manager: "The accountant constantly complains about the
EXCESSES. He uses ONE overhead rate only. I've got five product groups
on site and a service activity requiring a range of manufacturing activities
and respective process organisation encompassing:
- Labour intensive winding unit assembly
- CNC processing
- Jobbing shop and so forth

The accountant applies just ONE rate and expects me to be accountable for
the EXCESSES"!

COMPANY B

Engineering Manager: "I can get almost every submission, within reason, through the system. The accountant does not stand a chance. A good engineering statement, submitted at the right time during the year; offering a payback of say 13 months, NO PROBLEM AT ALL! I make sure it crosses two financial years - the accountant has no way of verifying subsequent performance. He applies blanket rates; shopfloor production and performance is not recorded by machine or even process. Its absolutely great - you can't fail, use your common sense and you're home and dry. Just provide a statement of factory economies and put your power into the Engineering debate".

COMPANY C

Engineering Manager: "The Finance Director stops all development and progress in this business - he sets impossible levels of Return on Investment and Payback periods. He won't even tell me the levels required - returns nor payback. He simply reports that the project has been rejected.

In addition, there is no point at all in talking to him about the advantages and benefits resulting from:

- Reduced leadtimes

- CAD/CAM support facilities

- 'Just-in-Time' materials flow systems INTO and THROUGH the business.

Unless we can establish:

i)   Qualifying criteria, for example,
      the Ford certificate for Quality Control Systems.

ii)   Order-winning criteria,
      reduction of costs; availability when required, etc

We will not even be allowed to quote and to win orders. The accountant doesn't even understand what I'm talking about".

COMPANY D (MAKE/BUY DECISIONS)

Finance Director: "I prefer to control on a standard cost basis which excludes overheads. Overheads on this site are almost entirely fixed - there's no point in charging all the overheads back to the products.

The process of make/buy compares the suppliers quoted price against the standard cost (material and labour costs only) and provided savings of at least 15% (suppliers quote must be at least 15% below the standard cost) are evident we will put the work out.

We do have a problem in that it appears we are extremely competitive in the Press Shop, but overall the product group operates at a large loss".

THE ENGINEER HAS A CASE TO ANSWER

Mismatch of key manufacturing task to order-winning criteria.

Under-utilised dedicated plant.

Inflexible plant with excessive change over penalties required to manufacture a wide range of small batches of products.

Capital expenditure has been authorised and yet the above problems are still apparent.

Ref "Manufacturing Strategy" by T J Hill.

This book in consideration of the need for Manufacturing Strategy clearly sites numerous cases revealing the mismatch of the manufacturing facilities to the needs of the business requirements.

Some relevant aspects for a particular company	*Typical characteristics of process choice*		
	jobbing, unit one-off	*batch*	*line*
Product range	High diversity		Standard
Customer order size	One-off		Large
Volume of operations	One-off		High
Degree of product change accommodated	High		Low
Make-to-stock	No		Yes
Ability of operations to cope with new developments	High		Low
Dominant performance criterion	Delivery/ quality		Price
What does the company sell?	Capability		Products
Set-ups { number	Many		Few
Set-ups { expense	Inexpensive		Expensive
Flexibility of the production process	Flexible		Inflexible
Process time	Long		Short
Amount of capital investment	Low		High

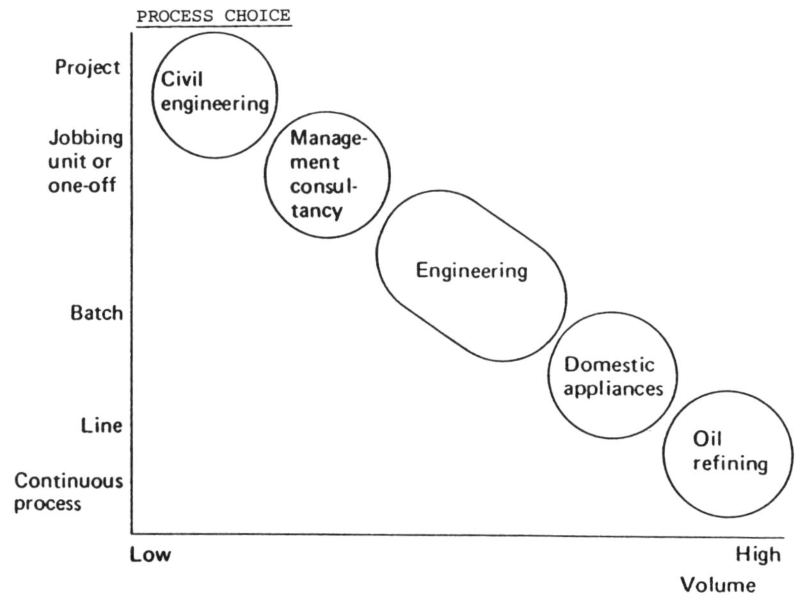

INVESTMENT AND COST IMPLICATIONS OF PROCESS CHOICE

Investment and cost	*Typical characteristics of process choices*				
	project	*jobbing, unit or one-off*	*batch*	*line*	*continuous process*
Amount of capital investment	Low/high ➡ Low ──────────➤				High
Economies of scale	Few	None ──────────➤			High
Level of inventory — components and raw materials	As required ──────────────➤				Planned with safety stocks
Level of inventory — work-in-progress	High	High	Very high	Low	Low
Level of inventory — finished goods	Low ──────────────────➤				High
Operations — material	Low/high ➡ Low ────────➤				High
Operations — labour	High ──────────────────➤				Low
Opportunity to decrease operation's cost	Low ──────────────────➤				High

© T J HILL 1983

MANAGEMENT ACCOUNTING PROCEDURES AND THE RELATED FINANCIAL SYSTEMS

"The problems and related problem statements have been extensively documented over the last few years. The most widely quoted statement:

"The Evaluation of Management Accounting"
by Robert S Kaplan
(Harvard Business Papers, Accounting Review July 1984)

The paper reveals the lack of any real accounting developments over the last 50 to 60 years, reviews a number of cases from which the issues and limitations of accounting practices are apparent".

GENERAL MOTORS

Recently published RECORD LOSSES but none of the General Motors factories made a LOSS!

BUSINESS RESPONSIBILITY FOR THE PROBLEMS

The business requires primarily of the Finance function:

1. Historical financial reports:
    - Balance Sheets
    - Revenue Accounts
    - Source and Application of Funds.

2. Examination of the business performance based on 'internal' results and records.

3. Reports providing a detailed examination of revenue expenditure.

4. The periods reviewed trace short periods of operating the business eg monthly, and in some aspects, weekly.

5. The primary measurements of performance relate revenue results to the investment used eg Return on Investment.

6. Analyse expenditure in respect of
        Products ...... which have been manufactured.
        Processes ...... the result of previous investment.

BUSINESS REQUIREMENT

FOR RECORDING	FOR INVESTMENT

1. HISTORICAL

1. Projected FUTURE
   - investment
   - performance
   - products

2. INTERNAL

2. Examination of competitive (EXTERNAL) environment.

3. REVENUE EXPENDITURE

3. Analysis of additional investment in CAPITAL expenditure and working capital.

4. SHORT PERIODS

4. Examination of future structure covering a number of years ...... Long periods.

5. RETURN ON INVESTMENT CONTROLS

5. Interpretation of CASHFLOW controls
   - dcf techniques
   - IRR
   - Payback period

6. EXAMINES
   PRODUCTS - already sold
   PROCESSES - previously evaluated.

6. EXAMINES
   Products - to be made
   PROCESSES - not yet installed.

BUSINESS ALLOCATION OF ACCOUNTING RESOURCES

	REPORTING	DECISION-MAKING
Accountants	High	Low
DataBase	High	Low
Computer Systems	High	Low
Established Accounting Techs	High	Low
Audit	High	Low
Publication	High	Moderate to low
Application	Regularly	Rarely
Promotion Route	High	Low

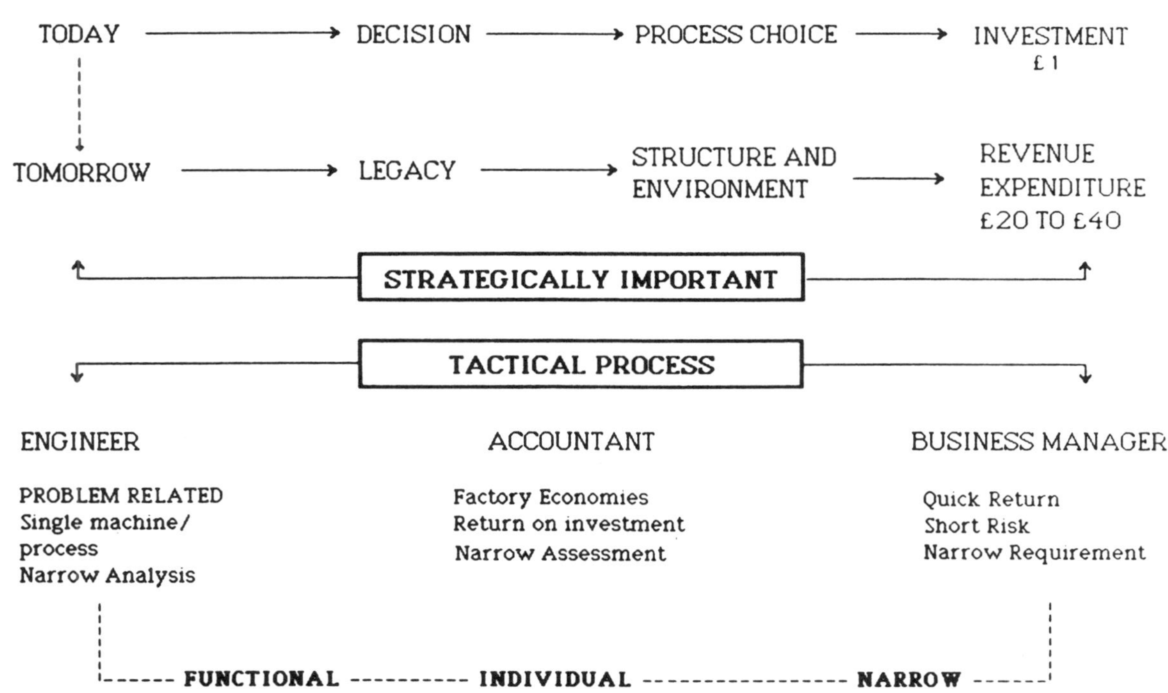

MISMATCH OF INVESTMENT APPRAISAL PROCESS
TO BUSINESS REQUIREMENT
TODAY
DECISION
PROCESS CHOICE
INVESTMENT
£ 1
TOMORROW
LEGACY
STRUCTURE AND ENVIRONMENT
REVENUE EXPENDITURE
£20 TO £40
STRATEGICALLY IMPORTANT
TACTICAL PROCESS
ENGINEER
ACCOUNTANT
BUSINESS MANAGER
PROBLEM RELATED
Single machine/ process
Narrow Analysis
Factory Economies
Return on investment
Narrow Assessment
Quick Return
Short Risk
Narrow Requirement
FUNCTIONAL
INDIVIDUAL
NARROW

EXAMINATION OF THE CONSEQUENCES OF INVESTMENT

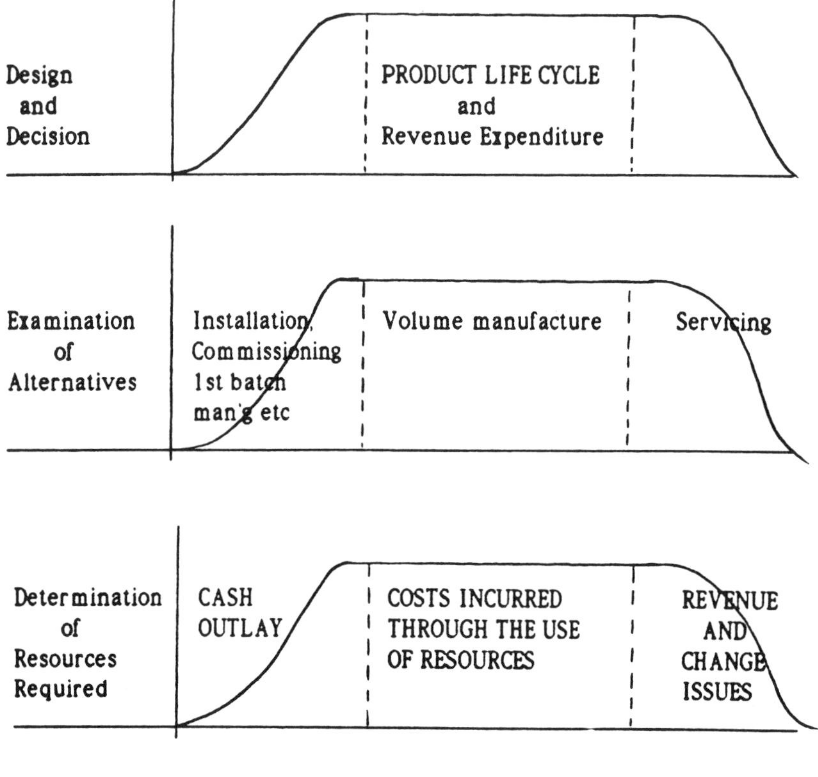

LEGACIES OF ENGINEERING DECISION

- Business and Investment Risks

- Structure and Process

- Revenue Expenditure

CONTINUUM: PROCESS CHOICE

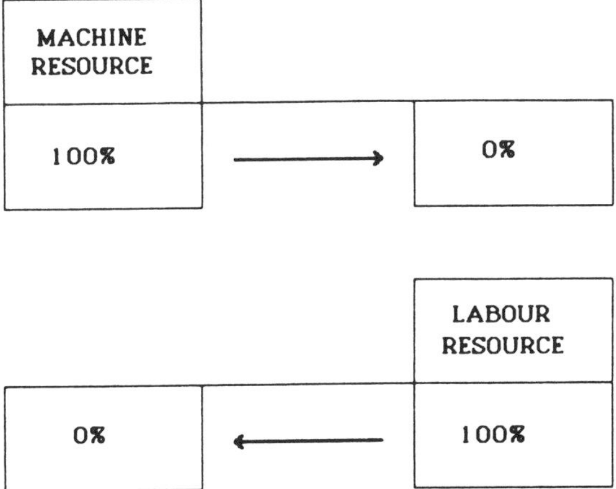

PROCESS CHOICE: PRIMARY RESOURCE

CONSEQUENCES

REFERRED RESOURCE	Business Risk	Structure - organisation support	Revenue Expenditure
MACHINE	"Upfront" Cash Commitment "Release" Disposal Cost	Works Engineers Setters Materials Handling Tool Maintenance	Power Maintenance Depreciation Tools
LABOUR	"Upfront" Recruitment and/ or Training Release Redundancy costs	Personnel Dept Wages Dept Training Dept Welfare: Medical Health & Safety	National Ins Holiday pay Sick Pay Training Special clothing Canteen subsidy

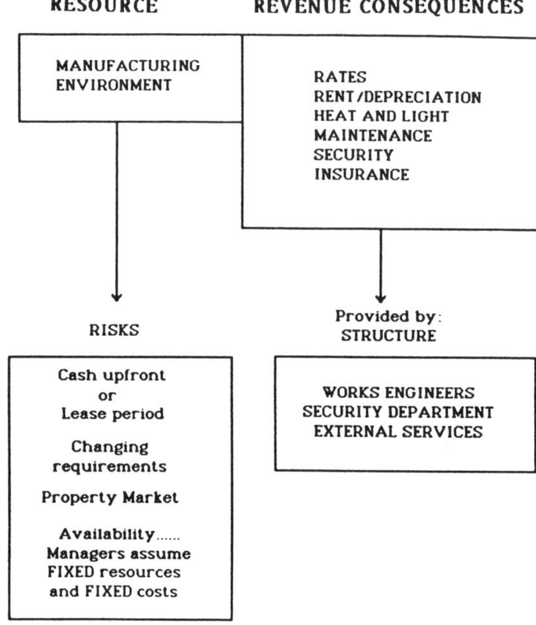
RESOURCE
REVENUE CONSEQUENCES
MANUFACTURING ENVIRONMENT
RATES
RENT/DEPRECIATION
HEAT AND LIGHT
MAINTENANCE
SECURITY
INSURANCE
RISKS
Provided by:
STRUCTURE
Cash upfront
or
Lease period
Changing requirements
Property Market
Availability......
Managers assume
FIXED resources
and FIXED costs
WORKS ENGINEERS
SECURITY DEPARTMENT
EXTERNAL SERVICES

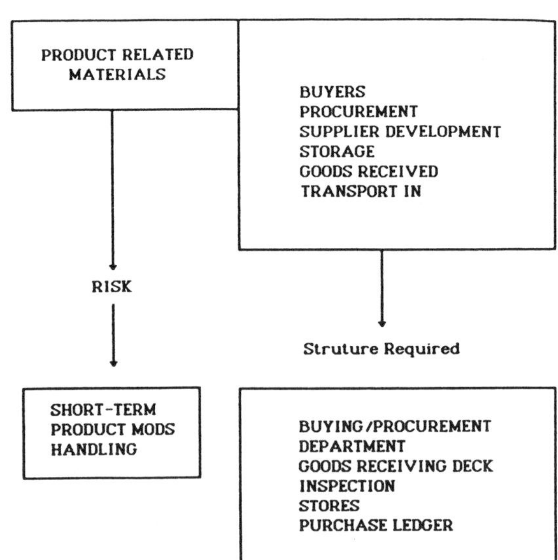
RESOURCE
REVENUE CONSEQUENCES
PRODUCT RELATED MATERIALS
BUYERS
PROCUREMENT
SUPPLIER DEVELOPMENT
STORAGE
GOODS RECEIVED
TRANSPORT IN
RISK
Struture Required
SHORT-TERM
PRODUCT MODS
HANDLING
BUYING/PROCUREMENT
DEPARTMENT
GOODS RECEIVING DECK
INSPECTION
STORES
PURCHASE LEDGER

ACCOUNTABILITY FOR REVENUE EXPENDITURE

- PROFIT CENTRE
- COST CENTRE
- APPROPRIATIONS/BUSINESS EXCESSES

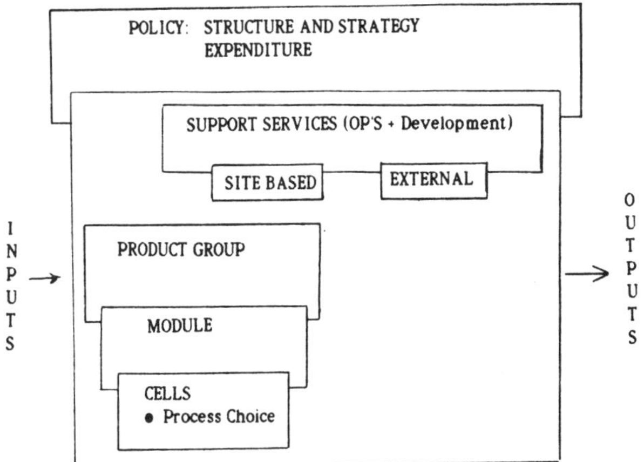

FINANCIAL CONTROL SYSTEMS

The control process is dependent upon the understanding of

WHY DOES THE BUSINESS INCUR REVENUE EXPENDITURE?

1. PROCESS CHOICE

    1.1 Machine based overheads.

    1.2 Labour cost and labour based overheads.

2. PRODUCT PORTFOLIO

    2.1 Material content.

    2.2 Material based overheads.

    2.3 Bought-out parts and resources.

3. MANUFACTURING STRUCTURE	3.1 Supervision
	3.2 Planning and Central Services.
	3.3 Materials movement.
	3.4 Buffer stores.
	3.5 Manufacturing Contracted services.
4. BUSINESS STRUCTURE	4.1 Policy
	4.2 Contracted Services

BOTTOM-UP ANALYSIS

.... tracing the material flow and determining the revenue expenditure
attributable to provide the preferred process choice and related
business structure.

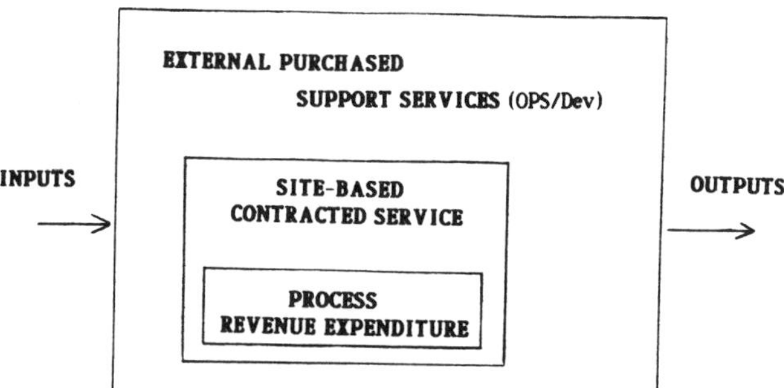

BOTTOM-UP ANALYSIS OF REVENUE EXPENDITURE BY PRODUCT, PROCESS AND STRUCTURE

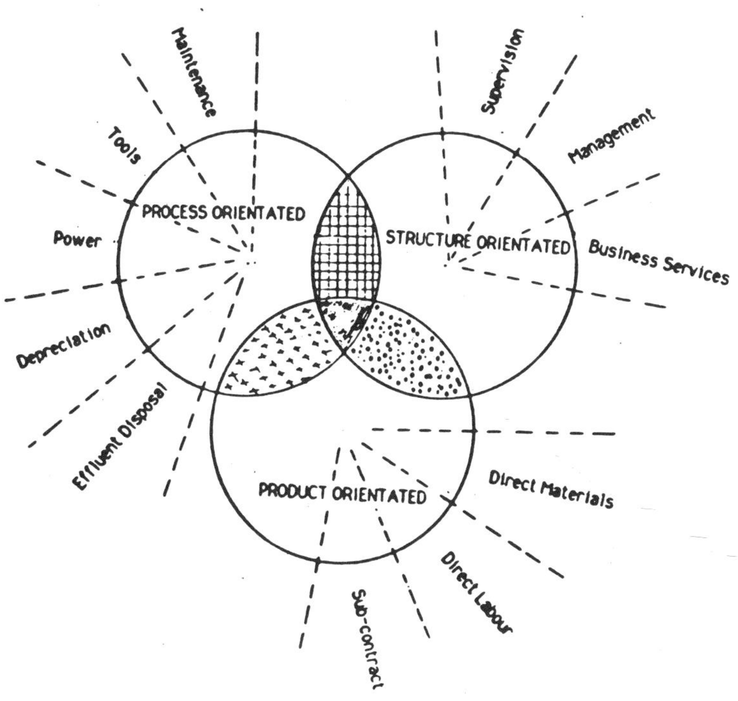

In a single product business the localisation and control of revenue expenditure could be more easily identified through the overlap of the three circles, indicating a high degree of dedication of structure to process and to the single product. However, in most manufacturing businesses the degree of dedication diminished and the complexity of accountability and control of revenue expenditure increases. In such businesses the traditional functional organisation should be questioned, and the localisation of services to the point of consumption linked to the re-definition of reporting lines should be pursued where possible. Certain services will remain on a support basis, either because of the economies of scale; the level of investment or perhaps the degree of specialist support required - therefore, policies must be established for the control and accountability of such services.

SUPPORT SERVICES: OPERATIONAL AND/OR DEVELOPMENT

TYPICAL ORGANISATION
"Functional"

ACCOUNTABLE ORGANISATION
"Contracted and/or Policy"

<table>
<tr><td>SERVICES</td><td>MANUFACTURING</td></tr>
<tr><td rowspan="4">F U C T I O N A L</td><td>PROD A</td></tr>
<tr><td>PROD B</td></tr>
<tr><td>PROD C</td></tr>
<tr><td>PROD D</td></tr>
</table>

<table>
<tr><td rowspan="4">C O N T R A C T   S E R V I C E S</td><td>PROD A AND IN-BUSINESS SERVICES</td></tr>
<tr><td>B</td></tr>
<tr><td>C</td></tr>
<tr><td>D</td></tr>
</table>

The planning and control process of the revenue expenditure required to support the structure of the "contracted services" would incorporate:

(i) bottom-up requirements - the evaluation of the needs (services to be provided) contracted by the manufacturing modules and/or other contract services.

(ii) top-down requirements - the evaluation of business policy and strategic development requirements.

The dual role of the support services should be clearly established to ensure the essential development of PRODUCTS; PROCESSES and PEOPLE to ensure the continuity of the business unit.

CURRENTLY: Services reporting to Manufacturing Manager

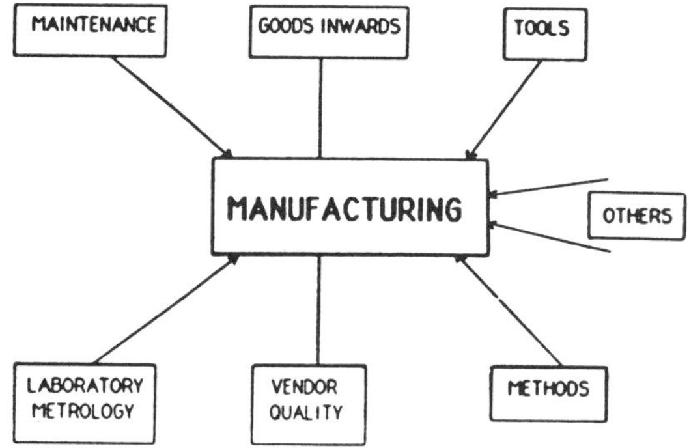

TRADITIONAL FUNCTIONAL SITE-BASED ORGANISATION

Proposed restructuring and localising of resources and control to the point required - where viable - through:-

1. Analysis of services currently provided eg Works Engineers Dept.

2. Analysis and restructuring based on Needs Analysis (using Input/Output techniques):

WORKS ENGINEERS ANALYSIS OF SERVICES

	PLANT Hours £	SERVICES Hours £	BUILDINGS Hours £	OTHER Hours £	TOTAL Hours £
ROUTINE					
BREAKDOWN MINOR MAJOR					
PLANNED					
RECONSTRUCTION					
SUB-CONTRACT					

FUNCTIONAL ANALYSIS

USING INPUT/OUTPUT ANALYSIS TECHNIQUE

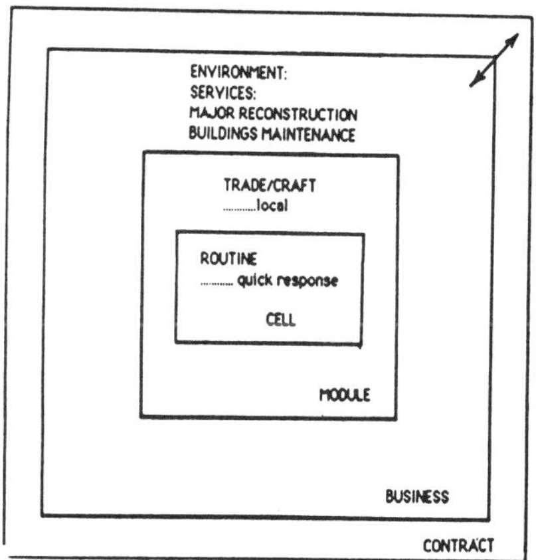

EXAMINATION OF BUSINESS SUPPORT SERVICES IN A MANUFACTURING BASED COMPANY

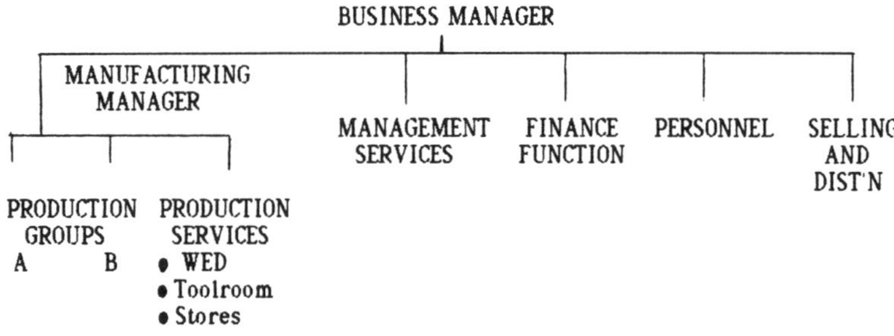

EXISTING SERVICES: organised on a traditional functional basis - controlled on a budget basis; difficult to determine effectiveness of revenue controls; budgets seen as a licence to spend:

- apportioned to cost centres or accepted as a fixed cost -
- difficult for cost centre to evaluate cost of service provided.

Application of bottom-up analysis to the Finance function:

CURRENT SERVICES PROVIDED through site-based function to 9 profit centres.	PROPOSED location of service.	RESTRUCTURE service required.
SITE ADMINISTRATOR	Contract Services (Site based)	Financial Ledgers; Financial Accounts; Cashiers; Cashbook; Wages preparation.
• Financial Accounts Sales Ledger Purchase Ledger Nominal Ledger	Profit Centre	Management Accounting Costing Estimating Materials
• Management Accounts		
• Costing		
• Estimating	Module	Wages booking
• Materials Control	Cell	
• Wages booking clerks		
• Cashiers		
• Wages		
98 EMPLOYEES		64 EMPLOYEES

A PROPOSED METHODOLOGY

- Reporting becomes routine and control by exception and POST
  AUDIT of previous decisions
  .... financial control requires a feedback loop.

- Decision-making is the result of analysis and evaluation of
  potential actions .... establish
    - Team approach
    - Database
    - Appropriate Techniques
    - Understanding of Financial consequences
    - Monitoring and feedback controls
    - Post Audit procedures and controls

- Change planning process from INFREQUENT TO FREQUENT
  eg 12 month moving plan.

- ESTABLISH CONSEQUENCES OF INVESTMENT.

- MODEL ALTERNATIVES.

- ESTABLISH ROUTINE REPORTING based on decision-making
  process.

- ESTABLISH ACCOUNTABILITY FOR RESOURCE MANAGEMENT

  AND

- COST RECOVERY ON LINES OF ACCOUNTABILITY.

CONTROL PROCESS

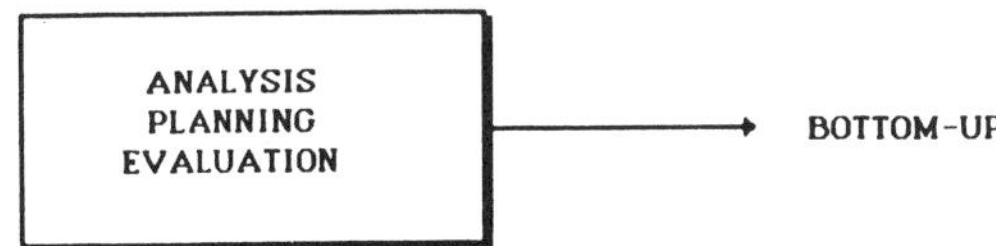

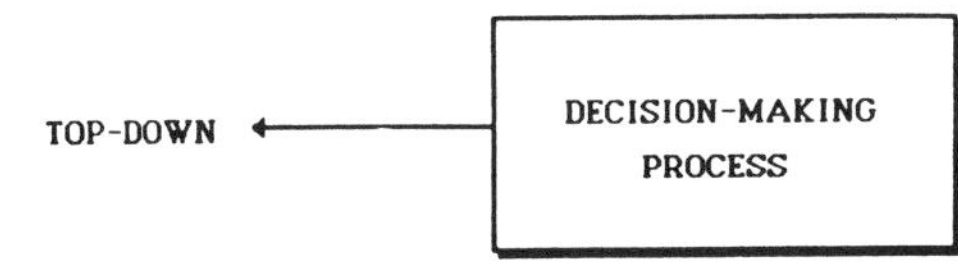

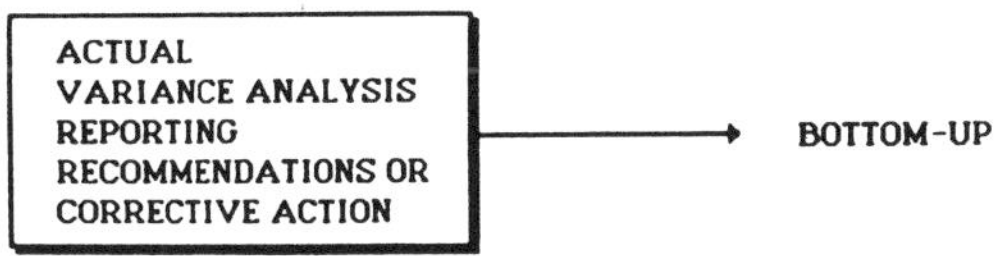

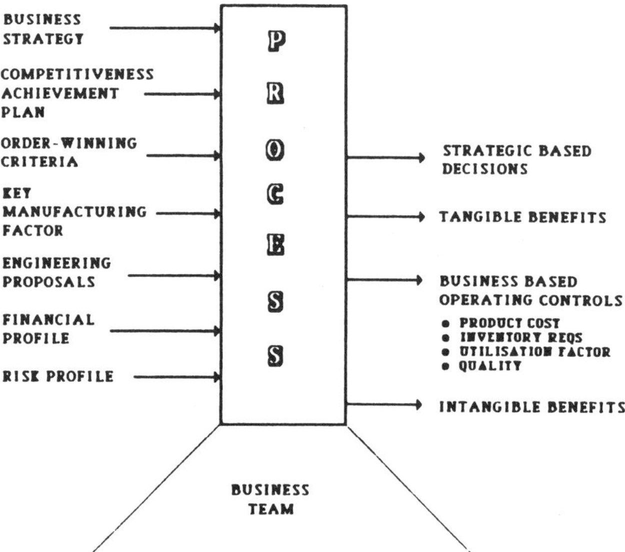

MACRO EXAMINATION OF INVESTMENT
ANALYSIS DECISION-MAKING PROCESS
BUSINESS STRATEGY
COMPETITIVENESS ACHIEVEMENT PLAN
ORDER-WINNING CRITERIA
KEY MANUFACTURING FACTOR
ENGINEERING PROPOSALS
FINANCIAL PROFILE
RISK PROFILE
PROCESS
STRATEGIC BASED DECISIONS
TANGIBLE BENEFITS
BUSINESS BASED OPERATING CONTROLS
PRODUCT COST
INVENTORY REQS
UTILISATION FACTOR
QUALITY
INTANGIBLE BENEFITS
BUSINESS TEAM

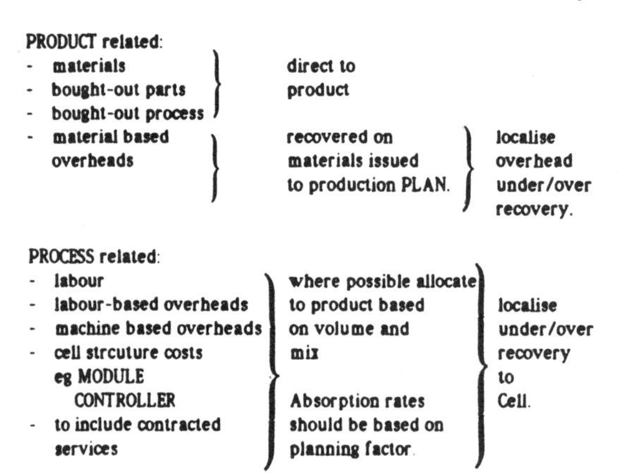

FINANCIAL CONTROL AND PRODUCT COSTING TECHNIQUES
PRODUCT related:
- materials
- bought-out parts
- bought-out process
direct to product
- material based overheads
recovered on materials issued to production PLAN.
localise overhead under/over recovery.
PROCESS related:
- labour
- labour-based overheads
- machine based overheads
- cell strcuture costs eg MODULE CONTROLLER
where possible allocate to product based on volume and mix
localise under/over recovery to Cell.
- to include contracted services
Absorption rates should be based on planning factor.

STRUCTURE related:

- Module - people and support services	absorption on completed	localise under/over recovery to module.
- Profit centre - people and contracted services and business controlled expenditure	PLANNED production	under/over recovery at profit centre level.
Policy related: structure; strategy development and business problems.	-	control by appropriation

The Engineer Must:

1. Model the revenue consequences:

    1.1 Alternative process choice.

    1.2 Alternative material-flow control systems.

    1.3 Manufacturing Environment requirements.

2. Examine and evaluate the impact of the process choices on:

    2.1 The existing business structure.

    2.2 The risk profile of the business.

3. In considering 1 and 2 above allow for the TIMING difference

    TODAY'S PROCESS CHOICE
    DETERMINES
    TOMORROW'S REVENUE EXPENDITURE

    ### ACCOUNTANTS must:

    1. Enter the appraisal procedure MUCH EARLIER.

    2. Provide the database for the modelling of resource based overheads.

3. Ensure that GIGO does not apply to the appraisal systems.

4. Audit:

   4.1 Capital investment requirements

   4.2 Working Capital requirements

5. Apply:

   5.1 Company-based accounting policies
        eg stock reserves, bad debts provisions, depreciation

   5.2 Cashflow conversion techniques from revenue
        statements

6. Must assist in the interpretation of the cashflow control
    techniques, and

7. Evaluate and interpret the RISK profile and consequent
    changes of the business structure.

ADVANCED MANUFACTURING TECHNOLOGY FOR POLYMER PROCESSING : THE JAPANESE EXPERIENCE

T. Naitoh*

Paper not available at
time of publication.

*Nissei Plastic Industries Co Ltd

ADVANCED MANUFACTURING TECHNOLOGY AND POLYMER PROCESSING—A MOTOR INDUSTRY VIEW

J.J. Lanfranchini*

New design requirements for passenger vehicles lead every
Automobile Manufacturer into new technological concepts
and an ever increasing consumption of plastic materials.
New materials and new methods of manufacture have required
P.S.A. to initiate an extensive internal development programme and the nature of some of this development work is
given. The lightweight and economical car concept needs
combinations of materials and processing methods which
could influence the polymer processing machinery designs
in the future.
An increased level of in-house manufacture and the
transfer of technology to selected suppliers have to be
considered in the body parts developments.

INTRODUCTION : <u>CURRENT TRENDS IN THE MOTOR INDUSTRY</u>

Motor-cars are currently undergoing major changes aimed at improved margins
and profits : industrial changes, social changes and technological changes,
involving new materials in particular.

This evolution in small or large-series vehicles is an on-going process, and
a policy of materials research and development is becoming progressively
established among car manufacturers, mainly as a result of continuous
improvements in plastics and composites. These now account of average for:
    72 kg per vehicle (8% in the PEUGEOT 205, 11% in the CITROEN BX)
    32% by volume of material used in the vehicle
    53% of raw material costs
    30% of hours worked
    20% of outside supplies for the manufacturer
    compared with
    220kg of polymers (rubber - 5%; paints, coatings and adhesives - 10%)

Concurrently, car manufacturers continue to be mainly concerned with :
- the fuel consumption of the vehicle.
- safety and comfort features.
- pollution control.
- corrosion resistance and durability.
- shaping innovations and renovation of styling, linked with cost-cutting,
  higher quality and generally improved productivity.

---

* Plastics and Composites Development Department (PEUGEOT/CITROEN/TALBOT)
  PSA "Design and Research" P.o. Box 16 - 92252 La Garenne-Colombes Cedex
  (France).

The car body is (after the engine/transmission assembly) one of the major
aspects in the search for greater fuel economy via the reduction of weight
and drag.

- The case for plastics and composites :

  The use of plastics and composites in cars and car bodies is justified by
  the specific adaptable or 'made to measure' properties of these materials,
  characterised by :

  - formulation and investment flexibility
  - energy absorption and vibration/noise damping capabilities
  - three-dimensional moldability
  - lightweight
  - corrosion resistance
  - ability to provide complex shapes and often to allow ease of repair.

  At the same time the overall costs of production of these new materials are,
  for a number of good reasons, decreasing constantly in comparison with
  those of steel.

  Thus, the current weight spread of plastics and composites in vehicles,
  averaging 63% for the interior and 15% for the body, should change
  substantially over the next decade to the benefit of detachable body
  components - still the major area of expansion for these materials -
  provided the price handicap can be successfully overcome.

- Technical and economic aspects of the development of plastics and composites

  One of the most significant factors affecting the development of body
  components will undoubtedly come about as a result of changes in shaping
  and in general styling of vehicles, using to full advantage the specific
  properties of these materials.

  The part to part replacement of steel components will only be practicable in
  terms of :
          - a direct financial gain relative to costs or investments
  and/or - an effective improvement of the product.

  The final cost to the car manufacturer and the effects on the overall
  price of the vehicle should be considered in terms of facility of
  installation of sub-assemblies, ease of internal access for personnel and
  tools in order to fit the body panels at the end of the assembly line,
  parts flow and parts stock controls, in particular.

  Price considerations usually favour the use of polymers for small and
  medium series, because of the lower investments involved.

  Moreover, the restyling of a small number of parts or sub-assemblies
  ensures - with a minimum of cost and timing constraints - greater
  flexibility to satisfy the requirements and preferences of the customer.

  The existence of various series and the wide range of models make this
  ability to cater rapidly and economically for the needs of the 'personalised'
  vehicle market a very attractive proposition.

Only the materials and processes used for major body components will be
discussed in this survey, despite the fact that significant technological
developments have also been achieved inside the vehicle and within the
engine compartment.

- Developments in car production and manufacture

It is commonly accepted nowadays that the major development of large-series
body components will be achieved through action in the following areas :

- the 'novelty' of the product (renovation of conventional styling)
- acceptable cost limits
- the conception of the product and of the materials
- technological progress at competitor level
- changes to the industrial set-up.

As progress is achieved in the use and processing of plastics and composites,
resulting in lower costs, profitability at series level will increase.

Thus, the existing pattern is likely to undergo substantial alterations,
involving :
- redrafting of car specifications
- changes in production requirements leading to a reduction of present
  high equipment costs (presses, paint lines, etc.)
- a technological breakthrough based on highly efficient materials and
  processes.

Automated production of vehicles, in order to be effective, necessitates
an adapted vehicle design. This entails reducing the number of components
and fitting sub-assemblies outside the main assembly line. These operations
can be facilitated by the use of synthetic materials.

Because of the present economic situation of the motor industry and the
frequent emergence of new models, direct developments on a large scale are
no longer possible without running major risks.

Consequently, any new solution involving technical or financial risks
should be evaluated :

- in a pilot plant, using production-scale facilities
- on experimental pre-series or small series based on large-series
  technology.

With innovation and pricing subjected to heavy pressure from competitors,
the car manufacturer should set in motion an active Research & Development
policy in close cooperation with outside partners.

ORGANISATION AND PROGRAMMES IMPLEMENTED WITHIN THE PSA GROUP

- Plastics and Composites Development Department :

The development of plastics and composites within the PSA Group has for the
past five years been the responsibility of a special Department, the
purpose of which is:

1) To develop materials and processes suited to large-series car production,
   combining reliability with consistent properties and low costs.

2) To mark out the future based on the vehicle, its definition and its
   evolution.

This task involves the identification and evaluation of materials and
technologies, followed by reporting and assistance to the various member-
company units engaged in the design and development of vehicles and
components, in order to validate the design of the plastic and composite
parts.

The Department examines the feasibility of new adaptations in car manufacture
by conducting research into the materials and processes involved, and then
assesses the effects of these innovations on the overall vehicle concept.

These requirements call for in-depth knowledge of the polymer market and of
the relevant processing requirements. This is achieved through in-house work
and by maintaining close relations with leading industrial concerns,
including suppliers of materials, molders (whether or not integrated),
from the beginning to the end of a project, and also through the assistance
of universities, research centres, etc. The Plastics and Composites
Technical Centre, equipped with pilot units, laboratories for quality
control of raw materials, formulations and component testing, develops
expertise to international standards in respect of the plastics accorded
priority within the PSA programmes.

In terms of the envisaged technologies, this expertise affects the overall
manufacturing process, i.e.

- component and tooling design
- processing equipment and procedures
- assembly and post-treatment
- checking and testing for adherence to specifications
- after-sales servicing and maintenance.

Thus, both the Department and the PSA Group have greatly benefited from the
experience acquired with vehicles such as the CITROEN BX and are now able
to draw certain conclusions regarding the materials and processes selected
for the manufacture of the vehicle. This in turn has led to re-thinking
and re-formulating the requirements of :

- the car specifications
- marketing and after-sales efforts
- the conception, design and development of parts and tools
- the design of processing and handling facilities
- the combinations of materials and technologies put to use
- data acquisition and processing for machine programming and process
  control based on the features of the actual material or compound formula.

One effect of this has been a quasi-redefinition of the entire design and
manufacturing process for the car or sub-assembly to be produced.

- <u>Typical examples of industrial achievement in the CITROEN BX</u>

From the start, the CITROEN BX was the only series-vehicle in which a
quarter of the total body area and 11% of the weight consisted of plastics
and composites, i.e. about 100 kg, including 43 kg of unsaturated polyester.

An important stage in the development of large-size body components has
been achieved in three areas :

1st Example

The front and rear bumpers (7kg) made from polypropylene
These match the general colour of the vehicle, and all 13 colours were
subjected to the florida exposure test. To ensure the success of this
venture, work was carried out on:

- **the polypropylene** : this is a natural sequenced block copolymer (with in-situ
EPR formation) developed from the new generation of high yield catalysts.
The system of UV and oxidation stabilisation has been optimised by
titration at different stages during processing, followed by (natural and
artificial accelerated) ageing of the components, concurrently with
pigmentation.

- **the processing press** : direct in-process colouring using a masterbatch was
  made possible by a suitable formulation for the pigment support matrix,
  in particular, but also by a special screw design ensuring plasticising
  and dispersal of pigments within a reasonable period of time (cycle time
  less than 90 seconds for an average thickness of 3.5mm). At the same
  time, the consumption of processing stabilisers was significantly
  reduced by careful monitoring of temperatures and screw work.

- **the component concept** : based on the 'soft-nose' concept, the bumper
  'skin' was directly secured at the top and bottom to the vehicle's front
  cross-member to form a hollow structure. The bumper passes standard
  testing at 4 km.p.h. within a conception combining maximum advantages both
  in terms of cost and in the utilisation of the properties of the material
  to minimise local distortion and stressing, in particular.

This achievement is a new and exemplary step for a car manufacturer. The
changeover from PP/EPDM mix to a copolymer designed for low-temperature
impact performance has made it possible to improve :

- the mechanical characteristics (modulus and scratch resistance)
- chemical resistance to gazole, in order to overcome problems of
  staining
- resistance to ageing: sequestration of stabilisers by the EPDM avoided,
  and added value of overall pigments accepted because of the lower cost
  of the polymer.

Overall, therefore, a cost saving on the PP/EPDM system has been secured
for the car manufacturer, coupled with a major improvement in quality. At
the same time, the molder and the material supplier have benefitted
from substantially lower margins over the past few years and also from
colour control facilities.

This example alone is sufficient to show the strategic importance of the
in-house and centralised work of a car manufacturer in this field, in order
to revise established practice and modify the commitment to specifications
governing the function, features of materials and processing requirements
which need to be taken into account as a whole.

At the same time, the methodology brought about by the implementation of
facilities for analysis, measurement, quality control and testing at the
various stages of the manufacturing process have facilitated automated production
on a heavy 3000T press (with 2 molded bumpers per cycle) and quick-changing
of molds by heat preconditioning outside the press, inter-platen transfer
and automatic clamping. The changing of colour gives less scrap by using
standard molding conditions available in the 13 colours.

2nd Example

The hatch back

This 8.3 kg component made of BMC polyester has been justified by the styling,
the integration of functions (3 pieces only), dimensional stability (thin
glass secured with adhesive) and the cost. The weight saving is 3 kg (with
the glazing) compared with a sheetmetal hatch back for which no design is
available but which would necessitate alterations to the styling.

The hatch back is produced by injection-molding at the CITROEN/RENNES and
MANDUCHER/POUANCE plants, located at a distance of 40 kms from one another,
based on a highly automated system of molding, finishing and 'specific
primary application' which is identical in both plants.

The rheology and reactivity of the compound, uniformity of mold temperatures,
rate of injection and titration have a very marked effect on the quality of
the components, both in terms of mechanical behaviour and appearance.

The advantages of this process are now clearly apparent. There are less than
0.8% rejects and less than 10% alterations to the primary product for the
molding shop with   90%-plus good components conveyed direct behind the
body finishing shop after 'sealer', lacquer and varnish.

Overall, the objectives of price and quality have been achieved. This
development, implemented concurrently in an integrated plant and with an outside
supplier, has shown the importance of the work methods and of the specific
facilities used in the motor industry to study, implement, develop and
achieve very rapidly a high level of industrialisation by transfer of
technology and close cooperation with a selected supplier within a field
which is still quite new.

3rd Example

The bonnet

The SMC polyester bonnet (9 kg) is justified by the need to overcome corrosion
and drag caused by the steep sloping front end, by resistance to minor impacts
and by the weight saving (5 kg compared with a steel bonnet).

This component is molded   by 2 outside companies, INOCAR and STRATINOR,
then delivered to CITROEN/RENNES coated with a specific primer in the form of
an in-mold coating.

Despite major efforts devoted to the design of parts, tooling, presses, IMC, post-treatment, etc., the process is at present a higher industrial risk than BMC injection molding, and productivity is lower.

In addition to the developments and robotisation currently being pursued (with important degrees of success in both companies), the tools and presses installed at the commencement of the project still remain, and it does not look as though any fundamental change will take place.

A new generation of concepts in terms of parts and tools, equipment, processes and formulations designed and developed by the PSA Technical Centre is progressing satisfactorily with a view to achieving a cost price (after in-line paintwork) approximating that of the sheetmetal used for this type of function.

DEVELOPMENTS IN MATERIALS AND PROCESSES : A CAR MANUFACTURER'S APPROACH (PSA)

The objectives of reducing the cost price of the vehicles (based on the same or higher performance features), the redefinition of assembly lines and post-treatment (robotisation), the wide variety of models and the rapid pace of new introductions to the market call for a thorough re-thinking of the overall concepts of the motor-car: in terms of shaping, design and manufacture using both conventional and new materials.

Plastics and composites could provide an answer to this general approach.

- Materials and basic technologies selected for body component developments :

In the medium term (the paintwork at 140°C), thermoplastic alloys, unsaturated polyesters and RRIM polyureas are being increasingly used for vertical components. For horizontal components (bonnet, roof, boot) unsaturated polyesters (BMC and SMC) will most likely continue to be used into the 1980's.

To achieve the best price/performance combination, the general specification is still essentially as follows:

- one molding cycle lasting from 1 to a maximum of 2 minutes, in order to match the vehicle output rate (integrated production)
- no secondary operations
- versatile and modular investments.

The related conventional technologies are of four types :

1) Thermoplastic injection molding :

In this highly industrialised process, the economic constraints continue to be the cost of raw materials for the techno-polymers and the heavy investments compared with the size of the components.

Short-term action is mainly concerned with process control. In the longer term, the polymers should desirably undergo changes aimed at reducing the price of the material, improving dimensional stability and reducing viscosity in the molten state.

2) <u>Injection molding of Thermosets (BMC)</u> :

These are essentially unsaturated polyesters with a thermoplastic
additive, glass fibre reinforcement and additional fillers. Utilisation
constraints have hitherto mainly been the impact resistance properties.
This material provides the most attractive function cost out of the
various body component materials in terms of the modulus/thermal
performance combination, provided the component is not subjected to
excessive impact stressing.

Improvements in this area cover not only the formulation but also the
preparation of the compound and processing, where major savings can be
expected.

3) <u>Compression of Thermosets (SMC)</u> :

These are also unsaturated polyesters with a thermoplastic additive,
glass fibre reinforcement and additional fillers.

The problems of industrial utilisation lie in process reliability and
variable production costs. For the body "skin" components it is
essential to develop :

- an in-mold coating formulated in keeping with the chosen motor-vehicle
  function.
- full control of the manufacturing process, from pre-impregnation to
  the finishing of components.
- a particularly efficient and optimised design of components and
  tooling, and development of molds.
lastly,
- the measuring, quality control and testing facilities essentially
  required to ascertain and judge the rheology of the material, the
  surface condition of the components, etc., as well as facilities for
  statistical process control. These facilities will also ensure
  improved development of formulations and of methods for the preparation
  of pre-impregnated components.

4) <u>Reactive molding by injection of reinforced materials (RRIM)</u> :

Based hitherto on polyurethane, the limit of this material lies in the
molding cycle. Significant improvements are provided by quick-acting
self-released amino systems.

The RIM polyureas are serious competitors because of their thermal
performance (180°C), self-released capabilities and molding cycles
approximating to one minute.

The RRIM low-pressure technology has its counterpart, for
structural components, in the "Resin Transfer Molding (RTM)" process
based on various resins, including polyesters, epoxies and acrylamates.

Likewise, the compression of reinforced thermoplastics by stamping is
developing as a counterpart of the SMC (Sheet Molding Compound) process.
Based on polypropylene and on polyamide, SMC molds can be used for the
production of semi-structural components either hidden from view or which
do not make demands in terms of appearance.

- <u>New material formulations and equipment</u> :

This rapid review of materials and technologies shows that :

- With the product conception being increasingly directed towards a
  choice of material, it is important that the problem to be solved
  should be clearly defined from the outset by means of a specification
  in which the constraints involved are stated in order of importance.

In this respect, the PSA Group's efforts are aimed at :

- The development and improvement of materials and formulations better
  adapted to the car production function and framework, backed by the
  relevant analysis and testing facilities.

- The construction and implementation of data processing facilities
  necessary for effective industrialisation and quality control,
  including an 'expert' data bank and CAD/CAM with
  molding control logic, based on a real-time data acquisition and
  processing system (Statistical process control) leading to servocontrol
  of the machine via its own programmable automation.

- The measuring methods and facilities required at each stage of the
  manufacturing process.

- The development and integration, within the Technical Centre pilot
  plants, of equipment providing facilities for the comprehensive testing
  of new materials and processes on a vehicle pre-series production
  scale.

- <u>Combinations of materials and processes</u> :

The developments achieved in overall functions outside the car body
show in particular that :

- the motor-vehicle application is technically valid and realistic,
- the cost is very often prohibitive in the simpler functions (wings,
  roof, bonnet, etc.).

Consequently, there seems to be a case for developing certain new
concepts of motor-vehicle products and associating the strong points of
certain families of materials and principles of implementation in order
to combine their advantages. At the same time, research is being
conducted into the direct preparation of semi-products and specific
formulations at processing equipment level, principally in injection
molding.

The initial investment is a multi-purpose modular-design horizontal press
of 900T equipped with a data acquisition system developed by PSA, capable
of accommodating 2 interchangeable 90° injection units for thermosets
('ZMC') or thermoplastics. It includes also an RRIM molding head on the
stationary platten side. Subsequent testing and re-thinking led to a
more sophisticated PSA project which resulted in a flexible processing
unit (also of multi-purpose modular design) comprising:

1) <u>A hydraulically-operated vertical clamping unit with a force of 1500
   tonnes</u>, capable of clamping the press tool under fully parallel
   conditions (using a real-time computer system) with speed and pressure
   monitoring and programming, data acquisition, etc. This initial "MULLER"
   unit will come into operation in 1986.

   Programming and speed capabilities will make it possible, from 1986, to
   carry out all the cycles required for operations as wide-ranging as
   reinforced thermoplastic stamping, compression and in-mold coating of
   thermosets SMC, BMC, DMC and resin transfer molding.

2) <u>A combined thermoset (BMC) or thermoplastic injection system</u>, involving
   simple replacement of the screw conveyor/casing/hopper assembly or
   loading device. The system is equipped with hydraulic control throughout
   and with an elevator-mounted frame for adjustments to suit the thickness
   of the molds.

   This unit (which will come into operation early in 1987) will also cater
   for a certain number of combinations of materials and technologies. The
   design work has been carried out in conjunction with the BUCHER Company.

3) <u>An R RI M injection system</u>, for reactive compounds, capable of being
   combined also with the previously mentioned injection unit to provide
   mixed products having specific properties. When used on its own, it can
   implement the RIM or RTM techniques (high-performance compounds) for
   structural parts.

4) <u>A continuous-plasticisation injection unit for thermoplastics</u>.
   Alternatively feeding 2 transfer units to the mold via the distributor, the
   plasticising system is a modulatable profile twin-screw extruder. When
   detached from the 2 transfer units it can, on being rotated through 90°,
   carry out the preparation of certain thermoplastic alloys or compounds.
   As with the combined injection system mentioned above, this unit is
   mounted on one of 2 elevators located at 90°, according to the axis of
   the press. A prototype machine has already undergone tests in connection
   with this project.

This flexible processing unit can produce, on a single-vehicle basis, all the
car body components relating to potentially acknowledged or industrialised
materials and technologies. In particular, it provides facilities for
establishing with greater accuracy the limits of production feasibility and
for seeking the best possible combinations with a view to achieving the
final cost/quality objective.

This efficient tool was difficult to manufacture at subcontractor level and
only part of its flexibility is likely to be made use of in the production
plant.

In addition to its technical and economic importance from the point of view
of developments and preparation for the future, it also allows quicker and
more effective training of car workers in the use of plastics and composites,
so as to develop simultaneously their attitude of mind.

<u>In conclusion</u> : The advent of plastics and composites car body components seems
inescapable, but only the car manufacturer can at the present time co-ordinate
the initial work of design and industrial development in view of the
importance and complexity of the subject. The setting up of a pilot plant
and the performance of tests on vehicles are necessary to confirm the
technical and economic choices and to ensure transfer of technology under
the best possible conditions.

The choice of integration at car manufacturer level, covering part of the production of components and tools, seems inevitable. For manufacturers and suppliers of materials, this provides evidence of a deep commitment in this field, an assurance for the future in the area of investment planning. At the same time it guarantees the achievement of an effective level of industrial development and productivity by benefitting from the transfer of technology relating to the motor industry. The plastics and composites industry is still highly multi-disciplinary in terms of the skills required. Car body components, because of their size and appearance requirements, should not be faced with the least deficiency throughout the entire manufacturing process.

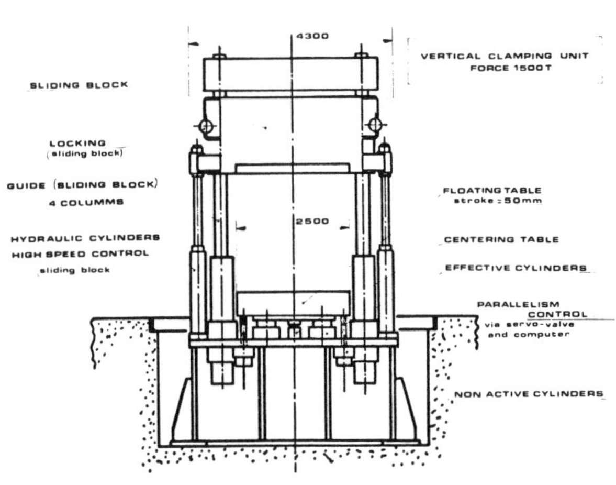

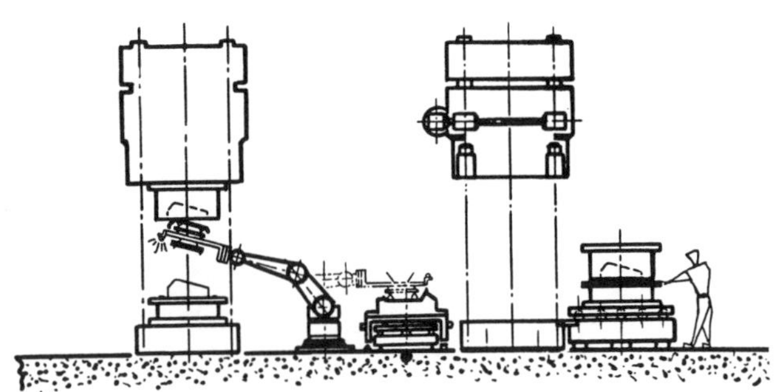

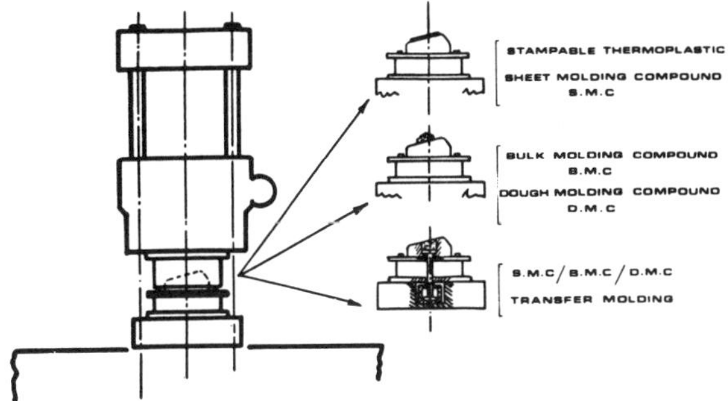
STAMPABLE THERMOPLASTIC
SHEET MOLDING COMPOUND
S.M.C
BULK MOLDING COMPOUND
B.M.C
DOUGH MOLDING COMPOUND
D.M.C
S.M.C / B.M.C / D.M.C
TRANSFER MOLDING
COMPRESSION MOLDING PROCESSES
OF THERMOPLASTICS AND THERMOSETS
(BASED ON THE VERTICAL CLAMPING UNIT)

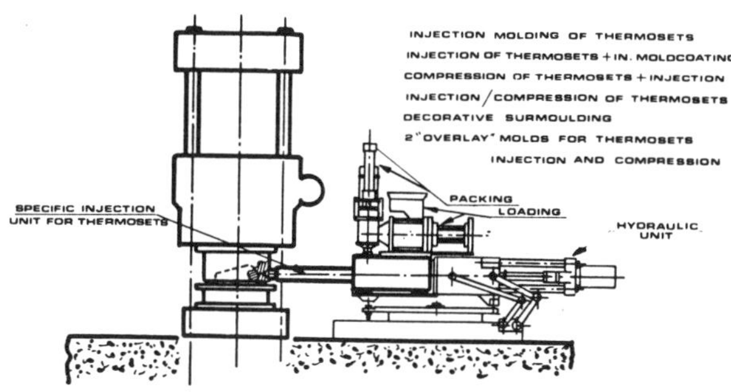
INJECTION MOLDING OF THERMOSETS
INJECTION OF THERMOSETS + IN. MOLDCOATING
COMPRESSION OF THERMOSETS + INJECTION
INJECTION / COMPRESSION OF THERMOSETS
DECORATIVE SURMOULDING
2 "OVERLAY" MOLDS FOR THERMOSETS
INJECTION AND COMPRESSION
SPECIFIC INJECTION
UNIT FOR THERMOSETS
PACKING
LOADING
HYDRAULIC
UNIT
INJECTION MOLDING OF THERMOSETS
VERTICAL CLAMPING UNIT + INJECTION UNIT (THERMOSETS)

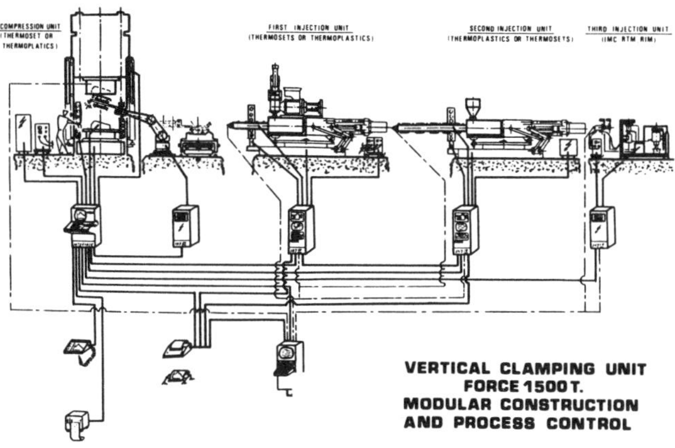
COMPRESSION UNIT
(THERMOSET OR
THERMOPLATICS)
FIRST INJECTION UNIT
(THERMOSETS OR THERMOPLASTICS)
SECOND INJECTION UNIT
(THERMOPLASTICS OR THERMOSETS)
THIRD INJECTION UNIT
(IMC RTM RIM)
VERTICAL CLAMPING UNIT
FORCE 1500 T.
MODULAR CONSTRUCTION
AND PROCESS CONTROL

IMPLEMENTATION OF EFFECTIVE MANUFACTURING SYSTEMS IN THE PLASTICS INDUSTRY—A SYSTEM STRUCTURE AND THE ASSOCIATED DECISIONS

A.K. Kochhar*

SUMMARY

In an effort to combat the stiff worldwide competition, companies in the plastics industry are being forced to consider the implementation of advanced manufacturing systems. Islands of automation or technologically advanced systems, in the form of individual sophisticated injection moulding machines or extruders, or a number of machines linked together, do not always lead to the desired levels of effectiveness. Indirect costs, high inventory levels and overheads remain major problems. Effective manufacturing systems can be implemented only by taking into account the detailed requirements and cost factors at the material flow, process control, production control and management control levels. This paper details the structure of an effective manfuacturing system, the parameters which should be considered at various levels, and the associated system decisions which must be made.

* Postgraduate School of Mechanical and Manufacturing Systems Engineering, University of Bradford, Bradford, West Yorkshire

CONTINUOUS INJECTION MOULDING: TECHNICAL FEASIBILITY AND INVESTMENT

Dr. Thomas Coppetti

Closed-loop injection moulding machine systems compensate short or long-time machine breakdowns themselves without the help of an operator. They are the base of all automation concepts. Now, also robot systems offer the possibility of a fully automatic change of moulds and plasticising units. This constitutes the fully automatic injection moulding shop. As is the case, such equipment proves a surprisingly good rentability, and as this paper proves, also improved product quality.

GENERAL INTRODUCTION

The essential feature of the fully automatic injection moulding plant is that change-over times can be drastically reduced. There are semi-automatic machines on the market, equipped with a high speed mould clamping mechanism which together with high speed data input and microprocessor controls already facilitate reductions in change-over times. One operator can change moulds and start up the new production. This can be done within a few minutes if the machine is closed-loop controlled.

In contrast to the semi-automatic machine, no operator is directly involved in changing moulds, plasticising units, plastic materials, handling devices and starting up procedures if the plant is completely automatic. Here, the job of the operator is to generally supervise all operations, prepare various components for subsequent jobs (e.g. moulds, plasticising units, raw material, demoulding robot) and intervene in case of a breakdown.

The purpose of the investigation described below was to prove that a completely automatic injection moulding line is capable of meeting the most exacting quality requirements with regard to moulded parts. This was done by chosing two very different mouldings which were produced completely automatically, i.e. computer controlled, alternately on the same injection moulding machine.

The tests were carried out on the NETSTAL automatic production line which was first demonstrated at the Düsseldorf Plastics Exhibition in October 1983 (K'83). 1. Although fully automatic changes could be demonstrated already then, there was, at that time, no exact quality assessment of parts immediately after the start of production. Moreover, the moulds used at K'83 were relatively simple whereas the moulds used in the present tests were more complex, some of them being hot runner moulds. The polymers also had to be pre-dried in different ways. Another complicating factor was that pure regrind had to be used for one production process. It was a known fact that both kinds of production process required a considerable amount of supervision when re-starting if carried out on conventional injection moulding machines.

These injection moulding trials on completely automatic machines were therefore of great interest, which centred on the number of cycles necessary to produce the first good quality part after an automatic change. The paper reports on these moulding trials and on the machines and equipment used.

Production aims

Products: The first product was a tap insert made from ABS containing 15% glass fibres. This part is produced using a six-impression three-plate mould. Pin gating with a heated sprue bush and two runners were used. The component must meet exacting quality requirements with regard to the internal teeth, which are checked with a test probe. Two important dimensions ($\emptyset$ 16.5 with the tolerances +0.1 and -0 and 46) were measured separately on each part (fig. 1).

The other product is a honeycomb unit made from glass fibre reinforced polycarbonate (fig. 2). This is made using a four-impression partial hot-runner mould. The parts are made in pairs, via a small runner, each with a film gate. This part must have extremely accurate dimensions. Not only must these honeycomb units fit together with extreme precision, but they must be able to firmly fix small cover plates with their frontal fastening cams.

The automatic injection moulding line

Description of plant and method of operation [1], [2], [3]: The completely automatic injection moulding line injection moulding machines, a pre-heating unit, a store for moulds and plasticising units, a computer controlled material feed system with interposed drying units, a superimposed control unit as well as the central production computer. The general scheme of the plant is shown in fig. 4.

The machine operator sets production and quality control data for the required production job into the central production computer, these data having been determined previously. For the fully automatic production changeover the central production computer issues commands, in good time, for changing-over mould, plasticising unit and raw material. The transport system, equipped with two grips, takes the required units from the store to the preheating station and then, after they have reached the correct operating temperature, to the injection moulding machine which just at that point has reached the required number of mouldings, thus concluding the previous moulding operation. Fig. 5 shows the transport system during the changing-over of a plasticising unit. Whilst the new unit (mould, plasticising unit and feed hopper) are being connected, the new process data are set by the central production computer into the injection moulding machine's control system. The new material is passed through the plasticising unit automatically for a short time and, after a break of less than 10 minutes, production can start again.

The power connections can be linked to heating and cooling circuits, high pressure oil (for core pullers), air and electrical current (mould heating), as well as to signal transfers of thermocouples and pressure transducers. To connect the mould on one injection moulding machine or preheating station, these power connection - which belong to both mould halves - are joined to their equivalent couplings on the machine. The leads between the power connections on the mould and the mould itself are installed only once, after which they remain firmly in place.

Results obtained

After each automatic production change-over the parts produced during the first 12 cycles were collected, marked and then measured. These measurements for each of the production cycles are shown in fig. 7 and 8. As fig. 7 shows, acceptable tolerances are achieved already with the second shot. From this second shot onwards, figures vary from the given required value 25.94 (+0/0.3) by only 0.04 mm absolute for all four cavities. The required figure of 1.5-H8 is also achieved from the second shot onwards and thereafter maintained. Individual values are separated by a maximum of 0.025 mm and the scatter of the control measurements, too, is in the range of less than 0.01 mm. Differences between the cavities, caused by the moulds themselves, are more substantial than the deviations due to the actual moulding process. Besides the above dimensional checks, the tear-off force of the small cover plates was also tested an found to be between 5 and 6 N, except for the first shot.

Before we discuss the quality assessment for the other moulding operation, let us see how the starting-up process is represented on the machines's VDU screen. Fig. 6 shows the display e.g. after 39 cycles in the production of the tap insert mentioned earlier. The parts counter indicates 216 pieces, i.e. bearing in mind that a six-impression mould is used, 36 cycles have already been counted since the fully automatic change-over. Only those cycles were counted in which the process variables were within the pre-programmed tolerance ranges. Besides these cycles, a total of three scrap cycles were registered, as can be seen in the "scrap cycle" column on the right. The production of scrap mouldings during these three cycles was due to the plasticising time being exceeded during those cycles, after automatic production started. All subsequent 36 cycles were within the pre-programmed tolerance range for the mean process variables which had likewise been pre-programmed by the central process computer during the change-over. As the figures in the "standard deviation" column show, the absolute scattering around the "mean values" is very slight, the standard deviation (3 o) for the injection time of 3.15 seconds being only 0.022 s. In other words, in 99.7% of all cycles, injection times varied by only around +0.022 s from the mean value.

There is another most interesting fact which is apparent from this "quality display page". A comparison between the mean values obtained for the first 36 cycles, displayed here, and the programmed "mean setpoints" shows that process data attained the level of subsequent continuous production already during the starting-up phase. This confirms that the starting-up phase of a feedback controlled machine is negligibly short. This, of course, applies only if the mould and plasticising unit have been sufficiently pre-heated. It should be stressed that this particular moulding operation was carried out using regrind.

In this test, an eye was kept on the production record during the entire starting-up phase, i.e. right from the first shot. As we have already said, no scrap mouldings were indicated after the third shot. Normally, the scrap rate display would have automatically resulted in the first three shots being separated from te rest. In this case, however, the parts produced during the first three cycles were duly marked and subjected to quality control together with the mouldings produced up to and including the twelth shot. As in the first example, it was possible also in the second example (tap insert) to maintain production tolerances during the first 12 cycles (fig. 8). Here, it can be seen that the setpoint tolerances $\emptyset$ 16.5 (+0.1/-0) are attained in most cases already with the first shot. It is evident, that exceeding a process parameter (first three shots) does not necessarily result in the production of scrap mouldings. The deviation from the set figure of 46.0 mm varies between -0.02 and +0.05 if one takes all the cavities into account. Checks on the internal teeth of the tap insert showed perfect results already from the first shot.

Let us now summarize the most important points:

These tests have shown that completely automatic production change-over is possible also with precision mouldings and can even lead to outstanding results. It is certainly surprising that the required dimensional tolerances were reached already after a few cycles, in some cases even with the first cycle.

Investment, Rentability

The NETSTAL automation system is available in various degrees of automation so that the described automation features can be fully obtained at one time, but also intermediate steps could be separately realized. These automation steps comprise the automatic mechanic-hydraulic mould clamping for step 1 (QMF, Quick Mould Fix), but also the automatic ejector coupling, as well as the automatic coupling of water, oil, electrical energy, electrical signals and pneumatic pressure for the mould, the automatic mould space adjustment, the automatic clamping stroke adjustment and the automatic closing force regulation for step 2, (QMC, Quick Mould Change) and finally the additional last automation step of an automatic mould plant for automatic handling of moulds and plasticising units including the automatic purging and the cleaning of the nozzles, the pre-heating station for moulds and plasticising units and the storage area for these devices (step 3, AMP, Automatic Mould Plant).

Rentability studies for such systems, valid for all of the equipment of the automatic moulding plant, show surprising and at the same time convincing pay back times, under the consideration of increased coverage of fixed costs resulting from weekend production, reduced stock of plastic products and savings in personnel costs. Judging by these conditions, it will not be long before the run for fully automatic and semi-automatic production plants in the injection moulding business will be a familiar sight.

Legends

Figure 1 - Detail drawing of tap insert
Figure 2 - Detail drawing of honeycomb unit
Figure 3 - General layout of an automatic plant
           1 central computer
           2 coordinating control
           3 material supply
           4 mould storage
           5 plasticising units
           6 transport system
           7 preheating station
           8 injection moulding machines
           9 demoulding robots
Figure 4 - Automatic plant: transport system with two plasticising units
Figure 5 - Diagram showing mould modifications for Netstal automatic plant
Figure 6 - Visual display of production control (quality supervision)
Figure 7 - Tolerance range for the honeycomb unit, made in a 4-impression mould from the first to the 12th shot.
Figure 8 - Tolerance range for the tap insert made on a 6-impression mould, from the first to the 12th shot (source: Netstal)

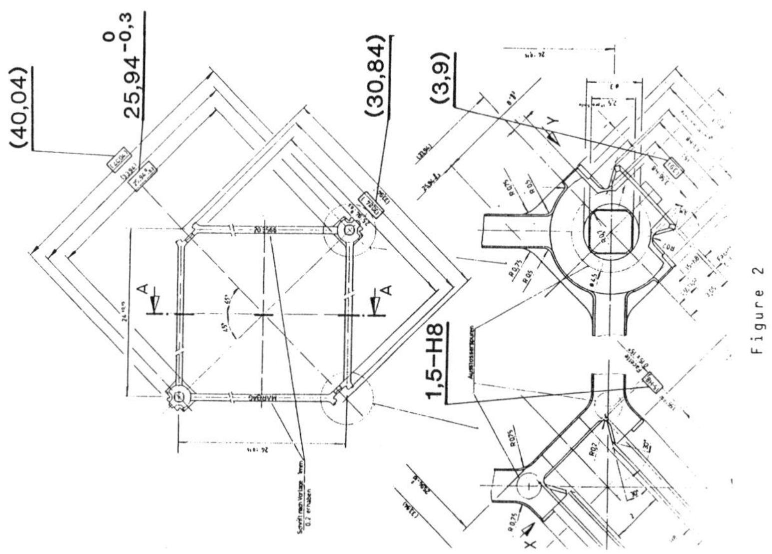

Figure 2

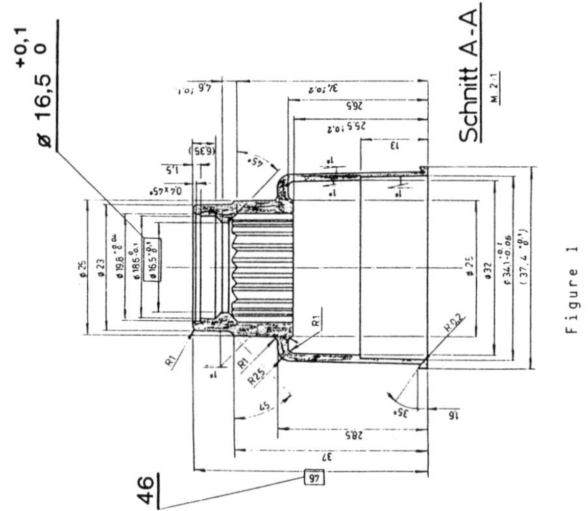

Figure 1

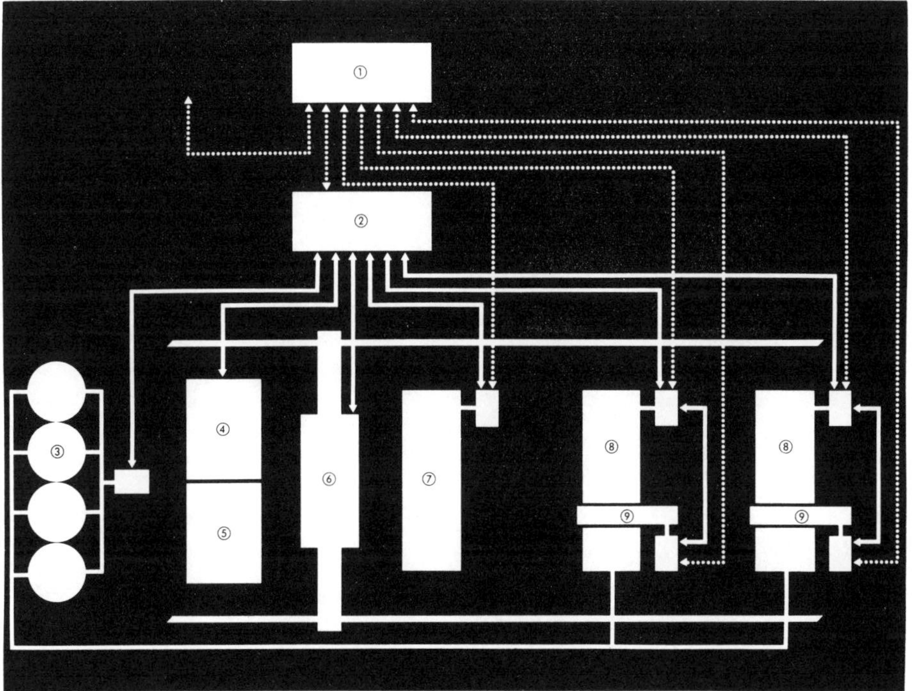

Figure 3

Figure 4

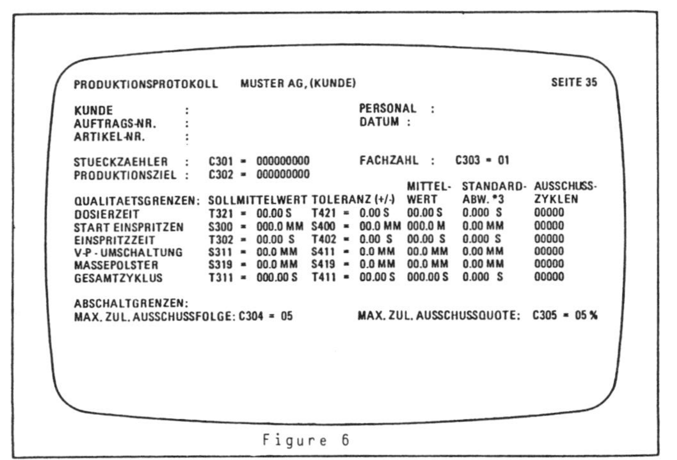

QUALITAETSGRENZEN:	SOLLMITTELWERT	TOLERANZ (+/-)	MITTEL-WERT	STANDARD-ABW. *3	AUSSCHUSS-ZYKLEN
DOSIERZEIT	T321 = 00.00 S	T421 = 0.00 S	00.00 S	0.000 S	00000
START EINSPRITZEN	S300 = 000.0 MM	S400 = 00.0 MM	000.0 M	0.00 MM	00000
EINSPRITZZEIT	T302 = 00.00 S	T402 = 0.00 S	00.00 S	0.000 S	00000
V-P - UMSCHALTUNG	S311 = 00.0 MM	S411 = 0.0 MM	00.0 MM	0.00 MM	00000
MASSEPOLSTER	S319 = 00.0 MM	S419 = 0.0 MM	00.0 MM	0.00 MM	00000
GESAMTZYKLUS	T311 = 000.00 S	T411 = 00.00 S	000.00 S	0.000 S	00000

Figure 6

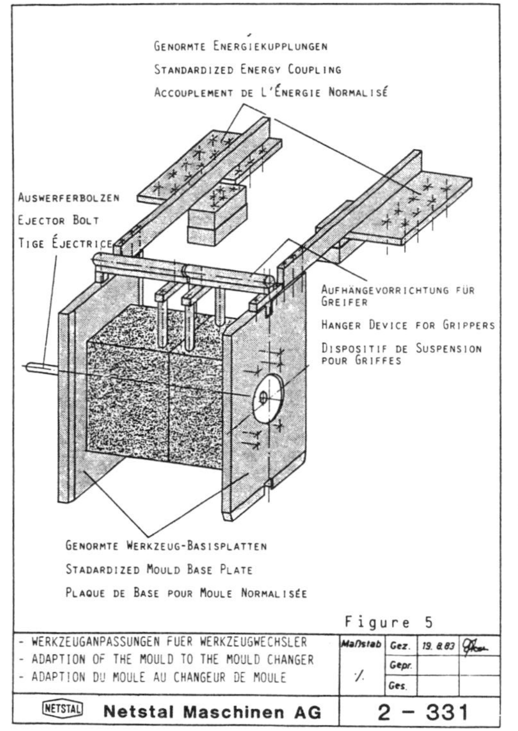

Figure 5

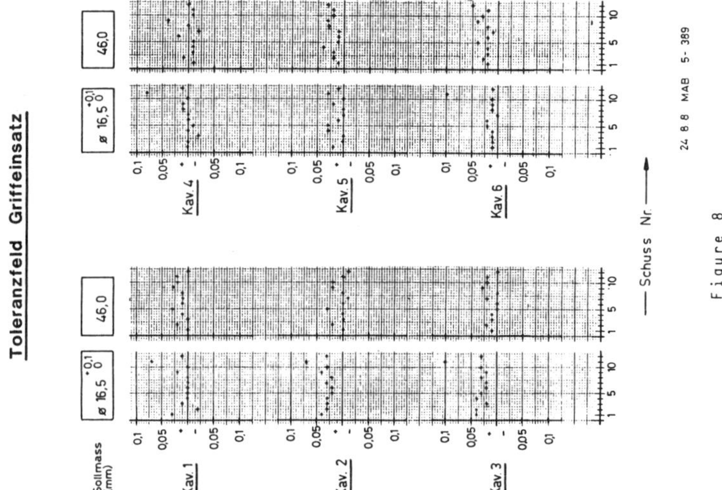

Figure 8

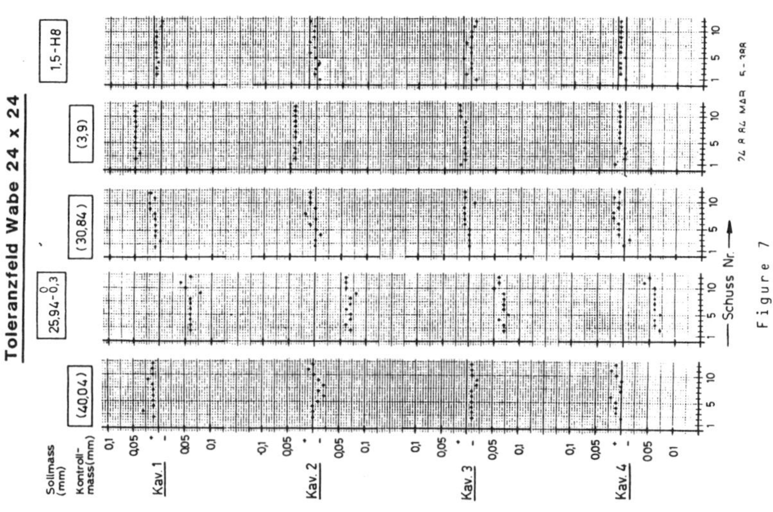

Figure 7

Part II

Computer Aided / Computer Integrated Manufacture

JUST-IN-TIME CONTROL FOR A POLYMER PLANT

I.D. Packman*

As customers insist upon smaller and more
frequent deliveries, the supplier's task becomes
more difficult to administer and he is pushed
towards holding higher stocks himself. The only
way out is to attempt to reduce batch sizes and
manufacture more in a Just-In-Time fashion.
This in itself can cause severe Production
Control administration problems if traditional
batch release logic is applied. A Simple Supply
System that minimises administration effort is
outlined.

THE PROBLEM

Quality Polymer Products Ltd were suppliers of a range of
components to the repetitive manufacturing industry (ie. motor
industry, washing machines, photocopiers, etc.) They were
going through a bad time.

 Benny, the Production Director, emerged from Tom Cat's Chief
Executive office suffering with shell shock. It was patently
obvious that the rest of the Board were looking for a scapegoat
- and Benny was it!

 Benny got back to the office, slumped into his chair and
pondered upon the points of Tom Cats acidic attack.

- Established customers were being lost because of poor
  quality and delivery performance. Sure they changed
  their delivery requirements at the last minute but that
  was the business that QPP were in - valuable long-term
  contracts with customer call of.

- Stocks were high. It was not only the money tied up
  (although that was bad enough) but the factory was a
  tip. Work in progress was everywhere, gangways were
  non-existent and it was embarrassing to take customers
  around.

*Manufacturing Systems Consultancy, Datasolve Limited

- Indirect costs were high. There were too many people
  in Production Control and the Stores.

- Quality reject costs were running at 10% - straight off
  the bottom line.

Benny decided to leave his office and visit the Production
Control Department in search of inspiration.

Production Control

QPP had recently installed a packaged Manufacturing Control
system which, through Bill of Materials, Inventory and Routing
modules, was intended to provide efficient costing and control
of shop orders. Savings had certainly been made in the
maintenance of standard costs but obviously Tom Cat's remarks
now cast doubt on any improvement associated with Production
and Inventory Control.

Benny strolled around the department and observed the
Production Control function:

- Customer call off quantities were loaded as orders to
  the system. Loading the moulding presses was regarded
  as the principle planning task with the assumption that
  finishing and pre-mould activities provided no
  bottleneck problems (fig. 1). Order batch quantities
  loaded to the moulding presses were identical to the
  call off quantities, provided that this represented a
  reasonable "run length" for the machines.

Questioning the Scheduler, Benny discovered that customer call
off quantities were getting smaller but more frequent. In many
cases these quantities were considered too small for efficient
running of the presses and consecutive call offs were being
combined to form a single order batch. Call off quantities
were therefore being produced before the customer wanted them.

Benny reasoned that not only was this likely to increase
stocks of some items, but it would also delay other orders
waiting for press capacity to become free.

- Three types of order were being released into QPP's
  manufacturing process (fig. 1): i) Moulding orders, to
  perform the moulding and finishing operations.
  ii) Extrusion orders, to convert compound material from
  sheet form into billets, defined by weight and shape.
  iii) Compound orders, to produce from raw material
  ingredients, sheets of polymer compound.

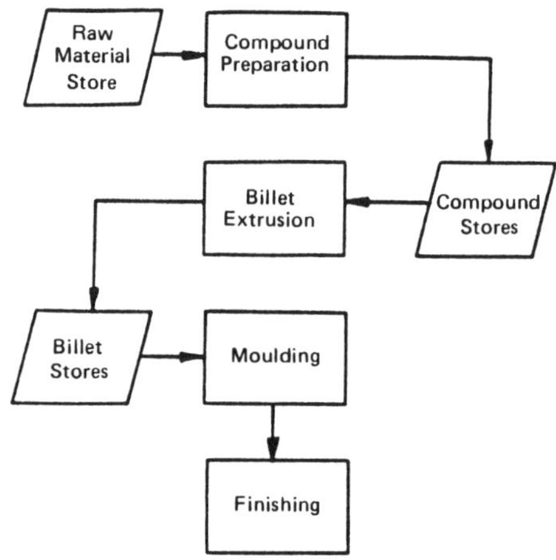

<u>Figure 1</u>   The Manufacturing Process

Extrusion and Compound orders tended to be of fixed quantity
and were triggered for release by the subsequent process
requiring them as material ie. as Moulding orders were released
(1 week before the planned start of Moulding), Billet stock
would be allocated to the order. A shortage of Billet stocks
would trigger Production Control to release an order for Billet
manufacture. In turn, a shortage of Compound stock for the
Billet order would trigger release for a Compound order (fig.
1).

- Changes to the manufacturing plan were a real problem
  once orders had been released. Shop floor documents
  would need to be ammended, material would need to be
  returned to stores.

Manufacturing plan commitment in terms of released orders was
generally 4/6 weeks before delivery. The increase in call off
frequency with smaller quantities was also leading to more
short-term program ammendments from customers - in many cases
well inside this 4/6 week horizon.

Production Control were faced with an increasing demand to
alter the manufacturing plan and handle all its attendant
administration problems. As large as thay were (according to
Tom Cat!), Production Control's resources were hopelessly
inadequate to react to such changes.

Benny returned to his office to ponder on his dilemma.

The underlying problem was that the moulding batch sizes were too high. All the other manufacturing processes were geared to the Moulding Plan.

- Finishing were attempting to process the large moulded batches in one go. Any attempt to split batches in Finishing was causing priority confusion and involved Production Control in what was essentially a Finishing replanning exercise.

- Billet and Compound manufacture was being triggered to support high stocks of both. Moulding order release allocated high stock quantities to one particular customer call off, causing a potential shortage for another customer call off.

- Scrap levels were high because moulding problems were only being discovered after finishing - and by then a large quantity had already been moulded.

By attacking and reducing mould change over times, Benny believed that he could reduce the batch quantities. This would not be easy and would require a high degree of commitment from his Moulding Supervisor and Production Engineers, but it could be done, the Japanese had proved it, by reviewing their mould handling methods, eliminating setting adjustments with the use of fixtures and by pre-heating moulds.

But if it was technically possible, how would the current Production Control system cope with a significant increase in the number of batches to be administered?

THE SOLUTION

As Benny was able to gradually reduce his moulding batch sizes he stopped using that part of his manufacturing control system associated with order batch release and consequent material allocation process.

He replaced this with a "Simple Supply System" based upon an approach known as KANBAN, eminating from the Toyota Motor Company in Japan.

The Simple Supply System

The Simple Supply System (SSS) is based upon two principles

- for each part, strict control of the quantity allowed to be held in a defined container

- for each part, strict control of the number of containers (and hence the number of components) allowed to exist as shop floor WIP.

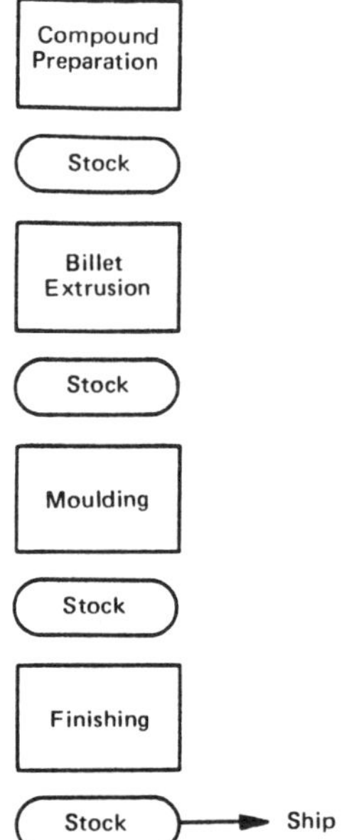

<u>Figure 2</u>   The Manufacturing Process Under SSS

The factory was viewed as four distinct processing areas —
Compound Preparation, Billet Extrusion, Moulding, Finishing.
Each had a formal area assigned to it to hold containers of
completed work (fig. 2).

For each component a level of stocks was planned to be held
after each process. Knowing the <u>average</u> call off rate and the
defined quantity to be held in a container, it was possible to
specify a number of containers to be held after each process,
which corresponded to a number of days WIP usage.

The number of WIP days (containers) planned to be held was
just enough to cover the period required to process a minimum
batch size (ideally one container).

Production Control Card Bank

Having defined the number of containers allowable as stock
after each process, control of this inventory was achieved by
insistence that each full container was identified by a CONTROL
CARD (fig. 3).

Production Control maintained a central bank of cards for
each product and process and as average usage rates altered,
either released additional cards or withdrew some from the shop
floor.

TYPE M	DEPT 1100	SER NO. 123
PART X10015 GEAR BOX COVER		
INSERT N/A QTY/INSERT N/A CONT C1 INSERT/CONT N/A CONTAINER QUANTITY 100		

Figure 3   Control Card

A simple microcomputer system maintained information on part average requirement rates, containers, container quantities and WIP stock requirement (in days usage). As requirement rates altered, the system advised upon action to release or withdraw cards from the shop floor.

Essentially Production Control administration was confined to:

-    maintenance of reasonably static micro system
     information

-    controlled release and withdrawal of cards to and from
     the shop floor as average requirement rates altered.

The Shop Floor Supply Cycle (fig. 4)

Stage 1 Stocks of finished product were held in containers in the Warehouse with a CONTROL CARD (F = Finished) attached. Goods were packed and shipped according to the Sales Schedule. As full containers were removed from the racking for quantities to be packed and shipped, the F-type CONTROL CARD was detached and sent back to the Finishing Department Supervisor.

In the sequence received, the Finishing Department Supervisor issues the card to his operators as authorisation to finish another container.

When another container has been finished the F-type CONTROL
CARD is attached to it by the operator, before being passed to
the Warehouse.

<u>Stage 2</u> To finish a container the Finishing Department
Operator pulls a container of moulded parts from the moulded
parts storage area. In doing so the M-type CONTROL CARD is
removed and sent back to the Moulding Department Supervisor.

<u>Stage 3</u> To mould a container quantity the operator may need to
draw on another full container of Billets (any partially used
container of Billets should be used up first), if so, he pulls
another container from the Billet storage area and detaches the
B-type CONTROL CARD which is then passed back to the Billet
Extrusion Supervisor.

The operator moulds a container quantity and attaches the M-
type CONTROL CARD to it, before passing it to the Moulded Parts
storage area.

This kind of gearing process continues back through Billet
and Compound manufacturing areas in the same way.

<u>Practical Considerations</u>

The principles thus described are the central core of the
Simple Supply System operation. Practical considerations must
be applied on top of this, however:

- if stock of moulded parts are not available for
  finishing say, then the F-type card would be passed
  immediately to the Moulding Supervisor. Such a card
  should never exist in his department unless there was a
  problem and he takes action accordingly. The moulded
  container would have both the M-type and F-type CONTROL
  CARDS attached to signify the need for immediate
  finishing. Similarly, for any other parts shortage
  through the different process areas

- initially it may not be practical to process only one
  container quantity at a time. In this case cards
  should be retained by the supervisors until further
  cards arrive for the part.

  It should be stressed, however, that this will increase
  the possibility of a shortage occurring and will
  require higher levels of stock to be planned for
  components having completed the process.

- Where customer requirements tend to fluctuate more
  wildly about an average, some additional stock may be
  required at all the process points. To a large extent,
  however, this can be accommodated by the system
  naturally circulating cards more frequently or more
  slowly. This same effect will occur when scrap causes
  an artificially higher requirement.

Management Style

A Simple Supply/Kanban type of Production Control system can
be seen to be a very effective way to administer work flow
through a plant operating Just-In-Time manufacturing. Control
of inventory is automatic to planned levels (no card - no
container!) and supply is continually fine tuned to
requirements.

The Simple Supply System makes little sense by itself, it is
a means to administer Just-In-Time manufacturing under minimal
inventory.

Traditionally, Production Control has tended to define
Inventory levels by default. The truth is that Inventory
levels depend upon batch size which in turn depends upon set-
up. The only people that can affect set-up times are Shop
Supervisors and Production Engineers. It is here that the
responsibility for setting Inventory levels (days supply)
should lie.

QPP recognised this and encouraged Shop Supervisors and
Engineers to help set initial work-in-progress levels for their
own process, with full knowledge of their own particular
inadequacies. They were then set targets for Inventory
reduction through set-up time reduction and judged accordingly.

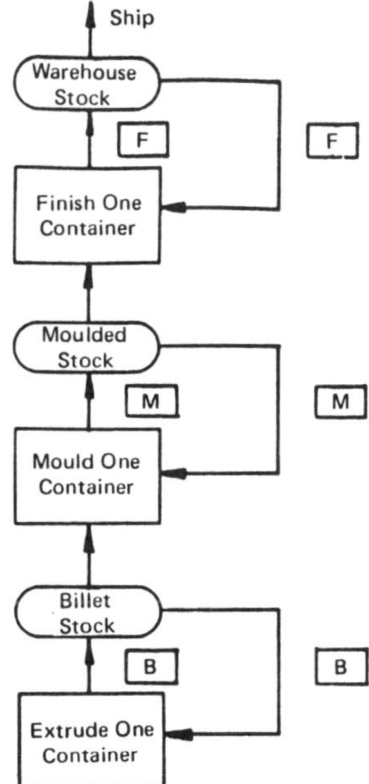

<u>Figure 4</u>   SSS Card Cycle

IMPLEMENTING COMPUTER AIDED PRODUCTION AND MATERIALS MANAGEMENT EN ROUTE TO COMPUTER INTEGRATED MANUFACTURING

R. Kallus*

Controlling a complete molding factory from one location is the plant manager's dream. This goal is reachable using todays hardware and the new upcoming software packages which integrate the various modules into one easy to use system.

At the present time we can watch some companies break their back in trying to change their entire operational procedure to "get there" with one big step. We believe in evolution, "walk before run", the plant has to install the basic system concept in place and then integrate one module at a time while controlling the work flow between the existing modules. Improving efficient and flexible production in small lot sizes. Getting the different jobs' production-schedule, parts orders, and tooling all arrive together to produce the desired unit. This setting demands data which must be collected from every point in and around the plant, and then made available in a useful form.

This technology is new to most plant managers who should first have to accept the overall system concept and then step into a modular implementation while keeping in mind the above stating goals of the total installation.

Let's have a look at a realistic achievable situation of a plastic plant in the 90's:

The typical plant will have machines made by more than one vendor, most of them will have a standard host interface to the common plant's network of information highway. The few machines which will not have the required interface will be augmented by an interface unit.

Production facility may be divided into three areas:

    * Material   * Manufacturing   * Tools

In order to properly manage a plastic (or any other) operation a production manager need a close watch on the obove areas.

The plant manager will have to add various other areas like:

    * Customers Orders   * Inventory   * Costs   * Labour

* Eurometrics bv, Netherlands

The plant integrated automation and information system will provide management with uptodate information in all levels of plant operation. For example, the production data will be fed in real-time to the central processing unit which keep updating its data base, all warehouses will be controlled by computer which manages the space allocation for incoming products, as well as finds and obtains any requested product to be fetched out of the warehouse according to shipping needs.

The best way to get a good feel for how such a plant will operate is to examine an order from the time it is received by the plant through the manufacturing process all the way back to the customer's hands.

When an order for a particular product is entered into the system, the computer will assign it to the best available machine. If the due date can not be fulfilled, or any other conflict has been detected, the user will be warned. At any given time the user may review all of the waiting orders, plant-capacity and machine loading, as well as the amount and type of the raw material required to produce those orders.

An order will be scheduled to run on a particular machine, and actually started by the operator from the machine site. Upon requesting the "Next Order" from the system few parallel operations will be done by the system:

*   The new set of control parameters will be sent to the machine in order to maintain the correct set-up points for the new order. The associated robot will receive its new program to handle the new product.

*   The currently mounted tool will be dismounted and moved into the tool room (into its pre-allocated space), the new tool will be fetched, moved into the machine site and mounted on the machine (all mold movements will be done by the system with no need for operator assistance).

*   The raw material routing will be adjusted to ensure that the required material will be pumped into the machine hopper, all additives will be prepared and the system will set the desired mixture.

The operator will oversee the entire operation but need not interfere with the process.

The machine starts to produce the product, there is on line inspection and scrap products are moved into the recycling operation, any secondary operation is done to the good products before they are put onto a container or a platform ready to be stored in the warehouse. The system knows the required warehouse's space needed for the product and prepares it ahead of time. During the operations the warehouse robot will periodically move the container/platform full of good products into their allocated space in the warehouse.

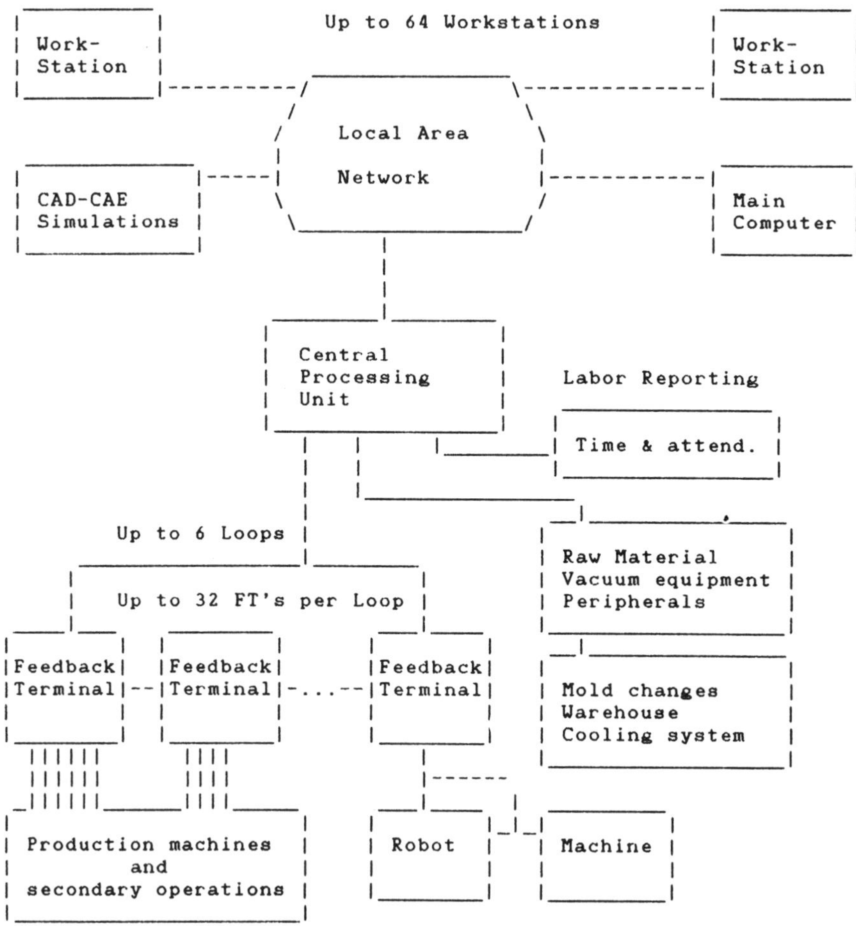

BLOCK DIAGRAM OF THE CIM CONCEPT

Up to 64 Workstations

Work-
Station

Work-
Station

Local Area
Network

CAD-CAE
Simulations

Main
Computer

Central
Processing
Unit

Labor Reporting

Time & attend.

Up to 6 Loops

Up to 32 FT's per Loop

Raw Material
Vacuum equipment
Peripherals

Feedback
Terminal

Feedback
Terminal

Feedback
Terminal

Mold changes
Warehouse
Cooling system

Production machines
and
secondary operations

Robot

Machine

Along the entire production route the system monitores the raw-material consumption, the machine performance, product inspection, secondary operations, robots activity and warehouse stocking. Every part of the operation compared with its expected standard and only exceptional situations, out-of tolerance operations are presented to the operators, who can take an immediate action to bring the operation within its standard performance.

The management information is continuously updated, thus every level of plant management may get a clear true picture of the situation, and base his decisions on a solid factual status.
The data is fully integrated to include all aspects of plant management:

    * Production - Last, Current, Future Planned.
    * Material   - On hand, Consumption rate, Future needs.
    * Tools   - Availability, Location, Performance compared.
    * Orders   - On hand, Schedule, In work, Est. Completion.
    * Labour   - Expected, Showed-up, Late arrival.
    * Good Prod. - Market needs, In house, In production.

When the sales people would like to schedule delivery of products they may examine the following data:

    * number of products in stock.
    * Expected completion time of current production.
    * expected production of currently scheduled orders.

A request may be submitted to the system to fetch any desired product from the warehouse. The robot will go and obtain the product from the warehouse and get it onto the loading duck.

The above description is not science fiction, all hardware components are currently available, while the software and overall system integration is couple of steps behind.

SYSTEM OVERVIEW

The system is led by customers' orders which entered into
the system and push all the other modules into operation.
There is a full duplex - two ways communication - between
the entry of the customer order, production schedule,
production order, real-time monitoring of production activity
with its various costs, on the spot performance analysis,
feedback (in financial term) to plant personnel, updating
inventory levels and manage purchasing of required material
etc.

The scope of this paper doesn't allow me to get into all
system's components in great details, but a quick overview
will help to understand how it all works.

First we should look at the company main Computer and
examine the architecture of a system that can cope with our
wide range of requirements and have hooks for future
expansion. The following figure depicts the structure of
the Industrial Management System (IMS). Some times called
"Manufacturing and Planning Control System".

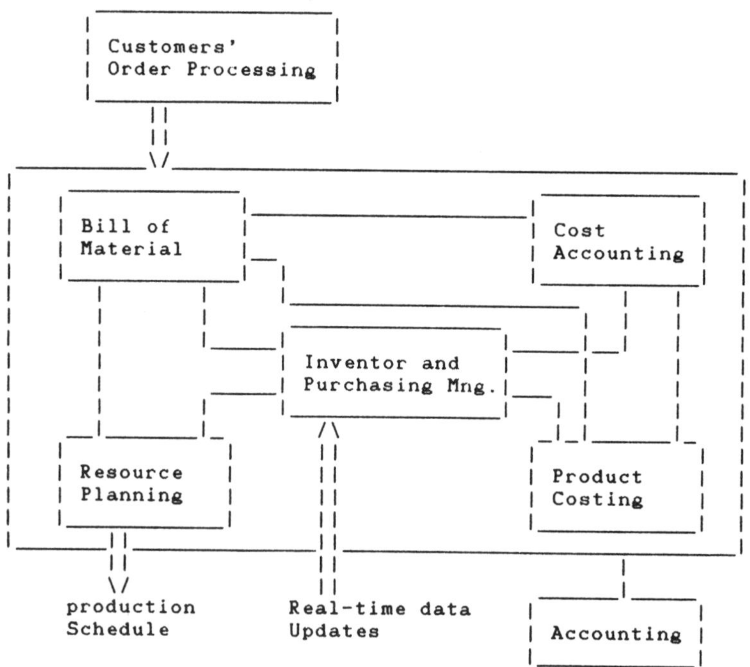

The main body of the IMS consist of five modules:

* Inventory and Purchasing Management

* Bill Of Material

* Resource Planning

* Cost Accounting

* Product Costing

Auxiliary interfaced with:
* Customer Order Processing

* Accounting System

* Real-time PMS

Throughout these modules the system maintains an on line customizing facility. This allows you to define many features of the end-user environment, eliminating the need for costly programming modifications.

Inventory & Purchasing Management

This module facilitates the management of inventory movement, balances and costs. Five fields (Standard, Average, Last, FIFO and Sales price) are maintained as the basis for inventory valuation, Cost Accounting and Product Costing. Several inventory balances (On-hand, allocated, reserved for customers, held, in-transit, and available to promise) are automatically maintained for a general inventory, by location and at a user-defined third level, such as lots or by production date. An extensive audit trail is provided for review and analysis.

Major Features & Functions:

* Multi warehouse
* User defined transactions
* Cycle counting
* Lot traceability
* End-of-period/ End-of-year processing
* Transaction history reporting
* Valuation reports
* Extended item descriptions
* Inactive inventory reporting
* 15 character, alphanumeric item codes, independent of
  item group codes
* Group technology
* Purchase order entry and printing
* Purchasing reports and inquiries

Bill Of Material

This module provides a tool for defining multiple level Bills of Material. An item bill may contain both material and operation components. Bill may be developed for buildable items, as well as 'planning bills' used for scheduling purposes. Single level and multiple level indented BOM explosions are provided.

The Bill Of Material module also uses the SYSTEM DEFINITION facility to define operations and work centers, which identify labor and machine components and costs, as well as productivity and maintenance norms. Bills defined in this module are processed by the inventory & purchasing management, Resource Planning and Cost Accounting modules. A powerful optional system feature is the automatic generations of certain inventory transactions based on the BOM definitions. For example, a production order transaction can be used to automatically generate inventory allocations (issues) or any or all raw materials or intermediate components used to produce (manufacture, assemble or blend) the parent product.

Another important feature of BOM module is the ability to duplicate components or operations in the same level of the bill, distinguishing between each by use of sequence (order) number, thereby creating routing definitions. Among other things, this allows for the printing of work orders by routing sequence.

Major Features & Functions:

* Unlimited number of levels
* Where-used reporting
* Bills may consist of materials and operations
* Loop detection
* Multiple versions of a bill, supporting models or options
* Effectivity dates
* BOM in any quantity or unit of measure
* Comments fields for alternative items or routings
* 'Same as, except for' copying
* Cumulative lead time calculations

Resource Planning

The Resource Planning module contains the functions relating to Demand Forecasting, Master Production Scheduling, Material Requirements Planning (MRP), and Capacity Planning.

Forecasts are entered by day, week, month, quarter or year,
as desired. This demand is then compared to present stocks
and scheduled receipts to produce a first level net demand,
time-phased master production schedule. After any review or
modification, the master production schedule is used for as
input to the MRP regeneration process, which uses the bills
defined in the BOM module to forecast net component and raw
material needs, again on a time-phased basis. Furthermore,
because Work Center and Operation data are maintained in the
BOM module, Capacity plan and machine loading are generated
along with reports showing net requirements for labor and
machines.

Major Features & Functions:

* Forecast can include any types of independent demand
  (sales, warehouse replenishment, planned purchases)
* User-defined time fences for forecast data and all
  subsequent processing and reporting
* Time-phased requirements projections
* Unbucketed transactions
* Simulation of plant's cash flow performance base on
  forecast sales, Current and planned machinery availability
* Capacity planning, machine loading reports by work center
* Interfaces with : Inventory & Purchasing Management
                    Real-time Production Management System
                    Customer Order Processing

Cost Accounting

The Cost Accounting module provides the ability to track
manufacturing costs by allowing you to create overhead cost
allocation models, used in conjunction with the Inventory
transaction records. As these models may represent actual
operating results or simulations, indirect costing and 'what
if' analyses are fast and simple. Five overhead allocation
methods are available for definition of allocation
percentages. Four user-defined cost center types may be
established in each company as the basis for cost analyses.
Furthermore, extensive selection screens enable you to
tailor the resultant financial reports or inquiry outputs to
your immediate requirements.

Major Features & Functions:

* Material usage, labor and machine time variance report
* Cost center reports by component and operation
* Profit/Loss reports by cost center type
* Cost roll-up reports using actual costs
* User defined reporting periods
* Table driven carrying cost calculations

Product Costing

The Product Costing module provides a convenient and simple
tool to calculate expected costs of selected items,
comparing them with alternative sales prices. It allows for
the display of direct labor and material costs together with
the respective indirect costs. The direct costs are obtained
from the Inventory and BOM modules. The indirect costs are
calculated based on pre-defined models established in the
Cost Accounting module. The analysis can be run in any
currency or unit of measure you select.

Profit/Loss percentages are displayed, and the various
component costs which comprise the total may be reviewed.
This module is an extremely useful tool for those involved
in price negotiations.

Major Features & Functions:

* Cost calculated for any quantity or unit of measure
* Cost analyses in any currency
* Summary or detail reporting
* Optional updating of standard costs
* Product costing based on models as defined in BOM
  (direct production costs may be calculated off-line by
  using various off-the-shelf package like SIMPOL, and
  entered into the BOM)

As a second step we should examine the Real-time Production
Management System depicted in Figure 2 and consist of the
following components:

    * Work Stations
    * Central Processing Unit
    * Feedback Terminals
    * Programmable Logic Controllers (PLC)
    * Time and Attendance recording terminals

BLOCK DIAGRAM OF THE IMPLEMENTED CIM
====================================

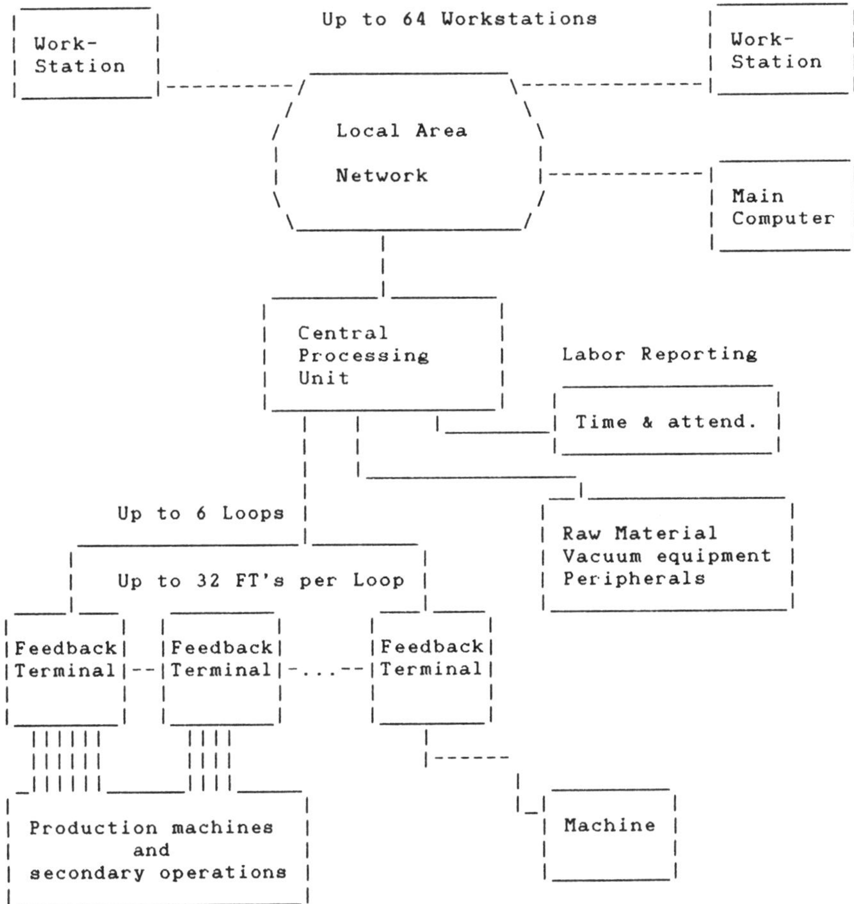

Work-Station

These are the primary interfaces between the plant
management and the activity in the plant. These are actually
personal computers (IBM PC's or compatible ones) which may
be used for many other application other than displaying
production management data.
The Work-station is connected as a node on the LAN and
communicates with the CU and the main plant computer as it
wishes. The local software knows how to obtain the user
requested information from the Central Unit and generate an
easy to read color coded screen. The color codes are very
important to enable a quick discrimination between the good
working machines/processors and the one who need attention.
The system promote the concept of "management by exception"
and helps the user to focus his attention on the out of
tolerance operations.

Major Features & Functions:

* Multi function, not dedicated for PMS
* Extremely user friendly, mainly used by function keys
* Color coded displays and graphic information
* Easy move from reviewing real-time PMS to the offline PMS
* Can access data on any storage device on the LAN
* Readily available sources for hardware components
* Can run many off the shelf third party software packages

Central Processing Unit (CU)

This component continuously collects the data from the
various plant locations, interpret and sort it into its data
base. On its other side it functions as a node on a Local
Area Network (LAN) and serves all inquiries by LAN users who
wish to review plant information.

At the End of each work period (shift) it generates a
summary of shift activity and makes it available for further
processing by the main plant computer or any other batch
oriented computer on the LAN. The batch processing will
incorporate the data collected during the shift into the
data base and updates all the necessary inventory,
production schedules and most of the reports.

Major Features & Functions

* No moving parts in the CU hardware
* Uninterrupted Power Supply
* Easy connection to "smart machines"
* Incorporation of Labor reporting system
* integration through PLC's with most plant operations

Feedback Terminals

These are the primary tools for two ways communication
between the machine, operator and the system. The Feedback
Terminals (FT's) shade the CU from all local special
interface requirements, industrial harsh environment and
relatively low skilled labor who operate the machines. From
its point of view the CU sees all machines and operations as
one type, the FT is the component that convert all I/O to
conform with the CU expectation. This configuration allows
you to mix any combination of new and old machines with
little or no effort. The FT collects automatic as well as
manual data entered by the operator or any other authorized
personnel. At the same time it provides real-time information
about the performance of the machine/operation in comparison
with the expected production performance.

Major Features & Functions:

* Four color lights to indicate machine performance
* LCD to display critical production indicators
* Keys for manual data entry
* 16 digital Input/Output ports
* 2 relays to control peripheral equipment
* work in a harsh industrial environment
* full duplex communication up to the CU and down to the
  machine controller
* optional data entry through magnetic card or bar-code
  reader
* Power can be supplied from a local main 85-240 volts AC,
  or through a central unregulated 18 volt DC.
* Battery back-up

Programmable Logic Controller (PLC)

The PLC's control various automatic and semiautomatic
operations in and around the plant. Some of the controlled
systems are:
>   + Water cooling systems.
>   + Raw material: Conveying, Mixing, Storage.
>   + Peripheral equipment and secondary operations.
>   + Movement of molds.
>   + Automatic warehouse operation

In addition to controlling the above operations the PLC's
send updated information concerned with the progress of the
process or operation to the Central Unit (CU). The CU uses
this information to update its knowledge of the plant
activity, compares it to expectation and generates alarms if
it finds parameters' values out of tolerance.
The Central Unit can download programs into each PLC and
change set-up parameters while the PLC is in operation.

PRACTICAL EXAMPLES OF COMPUTER AIDED PRODUCT DESIGN AND MATERIALS SELECTION

J. Trantina* and B. Rosvall#

Analysis of plastic parts using computers is carried out on a continuous basis by the authors and their CAED groups on both sides of the Atlantic. The paper describes how existing commercial software has been used, and how the results of such analysis have influenced part design and material choice. Further work is continuing on developmental software packages such as that to predict part distortion described here.

Product design using computers must satisfy the same criteria as that for any other design techniques. The product designer must consider whether:

- the product will fulfill the defined function for its life cycle.

- the product can be manufactured.

- the product costs and ultimate price will be acceptable in the market.

The power of the computer is extremely useful in analysis to optimise the design and material choice in terms of the above parameters. It is generally agreed that successful product design these days must take into account the manufacturing process, in terms of plastics materials the part shape or size may be varied in line with material rheological and/or shrinkage characteristics.

In this paper we shall look at computer aided design (CAD) and computer aided engineering (CAE) as it is used in the definition of part shape and size, and material choice.

We will also look at how analysis, made more practicable by the computing power available, may affect part design and material choice.

* General Electric Plastics, USA
# General Electric Plastics Bv, Netherlands

In order to restrict our field of reference somewhat we will only
consider product design and material selection in relation to
part performance in use, and during the injection of polymer melt
into the mould cavity and subsequent cooling. As such we will
only be considering injection moulded thermoplastic materials.
And whilst making a plea that all aspects of design and manufacture should be integrated, we will not consider runner or gating
systems for the product tooling.

<u>Hardware & Software</u>

General Electric Plastics have for many years offered, as part of
the technical backup for its engineering thermoplastic materials,
design advice and assistance for plastic parts. Initially based
on standard engineering formulae, simple predictions of part performance and material suitability could be made with the aid of a
slide rule and a data sheet. Within the space of the last ten
years we have seen a dramatic change from the above situation to
the point where we have a global network of computer aided design
work stations.

A programme of investment on CAED equipment was started in 1983.
In Europe we use stand alone GE-Calma units (currently S140 and
S120 models) complemented by a Digital VAX 11/785 (installed at
European HQ in Holland) which has a four megabyte memory.
Together both systems have a ten gigabyte filing capacity.

Terminals at the various technical centres throughout Europe,
including the UK, Germany, France and Italy are connected to the
VAX computer via dedicated data transmission lines. The main
frame in Holland is also connected to a similar main frame and
network in the United States.

The main software packages run are GE Calma "DDM", "Geomod", from
GE CAE Int. for part design and modelling in 3D, "ANSYS" from
Swanson Analysis Systems Inc. and "Moldflow" from Moldflow Ltd
for analysis of the designed part. Software development for
plastic materials is also undertaken by General Electric
Plastics.

Whilst we recognise that such an extensive CAED system is far
beyond the capability of the average plastics moulder or
toolmaker, in-house networks such as ours are being used by UK
industrial groups including their moulding operations.
Certainly, larger systems are being used for product design at
the major endusers, especially the automotive companies.

<u>Part Design</u>

When using a computer screen rather than paper on a drawing board
to define our product geometry we can do so in three basic ways:

<u>Wire Frame Modelling</u> - where we describe only the edges of the
product.

<u>Surface Modelling</u> - where the surfaces are mathematically defined
but not the intervening space.

Solid Modelling - where the whole volume is defined.

Wire frame models may be "meshed" with finite elements to allow definition of constituent surfaces and subsequent analysis to be carried out.

Surface models are of use in the preparation of NC tapes for computer aided manufacture (CAM).

Solid modelling systems allow us to look in detail at the form and function of parts and assemblies.

Often the analyses carried out on a part may not be performed on the same computer as the original design work. It is then necessary to consider data translation techniques from one system to another, such as IGES or VDA.

Material Selection

Given the main performance criteria which the part must satisfy, a number of computer data bases for plastic materials are available which allow the engineer, through a series of question and answer steps, to define which basic groups of materials may be used. Whilst the questions may act as prompts to avoid overlooking factors which may be important, these systems are often rather laborious and repetative.

Our own company is co-operating with M.I.T. Boston, MA, in building up and understanding the value of such a system.

The alternative is to have a computer based system which will allow one to call up data sheets on materials. To be of value these systems must be readily and cheaply accessible both for data retrieval and for updating detailed information. For this reason we are currently reprogramming our data base into a General Electric Plastics world-wide computer network, rather than using an independant network system, which proved both costly and inflexible.

Product Function

Where mechanical operations are concerned, and having the product design available on the computer screen, it is possible to simulate the functioning of the product under various conditions and to view it from any position. A process known as kinematics.

Furthermore, Finite Element Analysis (FEA) can be used to study the influence of mechanical load on the part. Whilst these techniques have been used for many years with metal parts and structures, widespread use with plastics materials is still relatively recent. The added complications due to the significant time, temperature and load dependancy of visco elastic materials like thermoplastics are often ignored in standard software packages.

We will use two examples here to illustrate the use of FEA with plastics.

The snap fit arm shown in Figure 1 was drawn as a wire frame and
meshed on the computer system. In this case since a single
operation of the snap finger on assembly was anticipated, and 80%
of yield stress was taken as the working limit, non linear behaviour was taken into account. With the material properties and
performance being defined, it was possible to calculate the area
and level of maximum stress. Figure 2.

Taking a more complex part, the electric motor housing for a
rotary ironing machine was being redesigned from an existing
diecast metal assembly to plastics. During the initial stages of
design the number of parts in the assembly was reduced significantly, Figure 3. In such a machine the motor is fixed at one
end and subjected to relatively high loads at the far end as
objects are forced under the roller. This was defined as a load
of 120N at one end of the motor applied perpendicular to the
shaft axis. The geometry of the housing was entered into the
computer and a data base established. Areas of maximum stress
were identified, Figure 4, and additional ribs and sections were
added, which could then be further analysed on the computer.

The final result showed a maximum deflection of 2.0mm where 5mm
was allowed, and a working stress of 17 N/mm^2. Thus the deflection was within the limit set and the stress level below the
recommended maximum stress level for the 10% glass filled polycarbonate chosen.

Information resulting from FEA can be used to see if a redesign
of a component is necessary, ie. additional ribbing changes in
wall sections etc.

Conversely reduced wall sections to decrease part weight and cost
may be possible in areas of low stress.

Material and grade selection are certainly influenced by the out-
come of such analysis, most often in terms of polymer type and
level of reinforcement to be used.

The maximum allowable stress level for the part can be influenced
by a number of factors including chemical environment, aging,
degree of crystalinity etc., and these must be taken into
account.

Ease of Manufacture

The most widely used computer based analytical tools which have
been specifically designed for use with thermoplastic materials,
are those software packages which analyse the flow of polymer
melts in tooling systems.

These can also be used to optimise product design in terms of
ease of manufacture, cost and performance.

Most useful in this respect are those software packages which
allow three dimensional treatment of the part. Given a 3D model
of the part on the computer this can then be "meshed", using a
number of finite elements to represent its surfaces, the gating
points chosen and material parameters defined. The results of
such a mould filling analysis may lead to changes in either part
design or material selection.

If the analysis shows that the part can simply not be filled under normal circumstances then thickening wall sections or choosing a lower viscosity variant of the material will be necessary. Conversely these programmes can be used to define the lowest practicable wall section to allow filling of the part and thus optimise raw material costs.

The ability to be able to display the flow fronts during filling of the cavity can be used to predict potential gas traps and weld lines. Where these are likely to occur in critical areas of the part, redesigning may overcome the potential problem. Apart from moving the gate position, using wall thickness variation, or ribbing as flow leaders can be considered.

Related to mould filling analysis is work currently being carried out by GE Plastics on software which allow prediction of part distortion immediately after moulding.

From the 3D mould filling study a flow gradient for the material in the cavity is established. This is then combined with a finite element stress analysis with transient thermal analysis to give the level of residual stresses caused by anisotropic shrinkage of the material, and the deformation of the part can be calculated.

Similarly the affect of variations in wall section can be studied with respect to shrinkage and deformation.

This software which was developed with the plastics group at General Electric is at the present time qualitative rather than quantitative, none the less it has proved to be a useful tool in predicting part warpage in glass filled semicrystaline materials.

Again one can relate the results obtained to either optimum material choice or to aspects of part design.

<u>Conclusions</u>

Work on our computer aided engineering and design system, which is used throughout the world to assist with design of plastic parts, has demonstrated that CAED can assist in the optimisation of product design and performance. Both finite element stress analysis and mould filling studies can be used to identify areas of potential weakness in the part which need addressing either by changes in part design or material. Conversely CAED analysis can help to avoid "over design" and unnecessary costs that this may involve.

The benefits of such analysis are evident but it must be remembered that costs of generating the information can also be significant both in equipment and time needed. Further availability of such software does not obviate the need for design experts for plastic materials, indeed the interpretation of the results needs skill and expertise.

To obtain the full benefit of CAED analysis on both part and tool design for plastic components, such analysis should take place at an early stage, ie, while the product is still under study by the enduser and not only when it reaches the moulder for immediate tooling production.

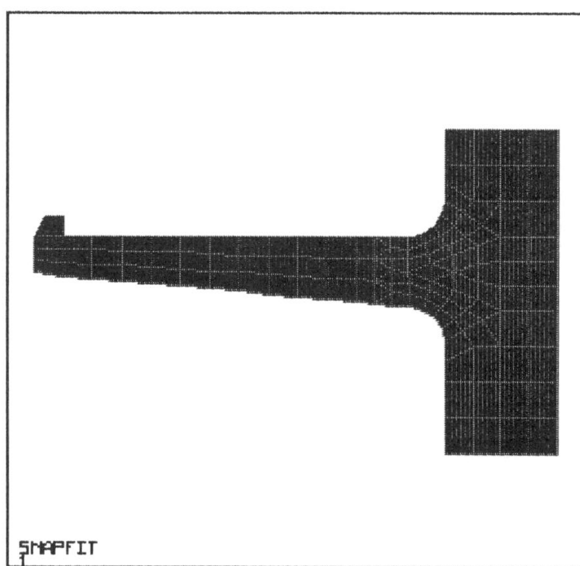

Figure 1. Snap fit arm wire frame model and "mesh"

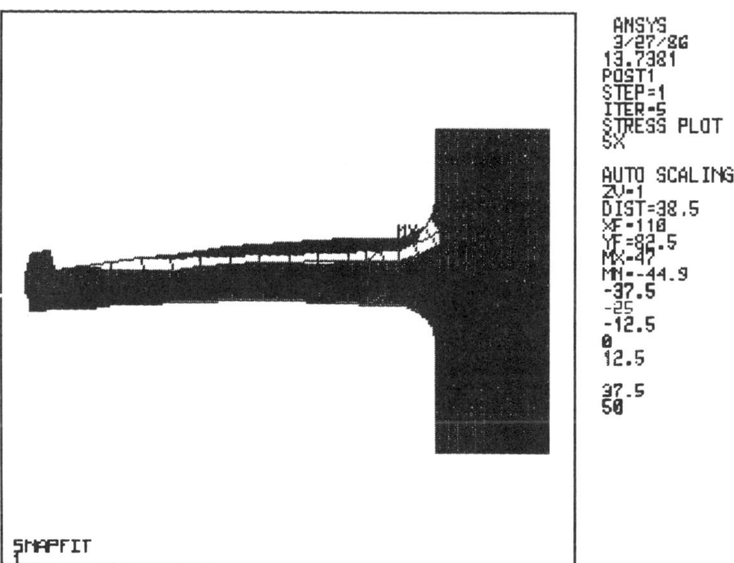

Figure 2. Snap fit arm graphical display of deflection and stress
           levels.

Figure 3.   Rotary press electric motor showing clam shell housing in 10% glass reinforced PC.

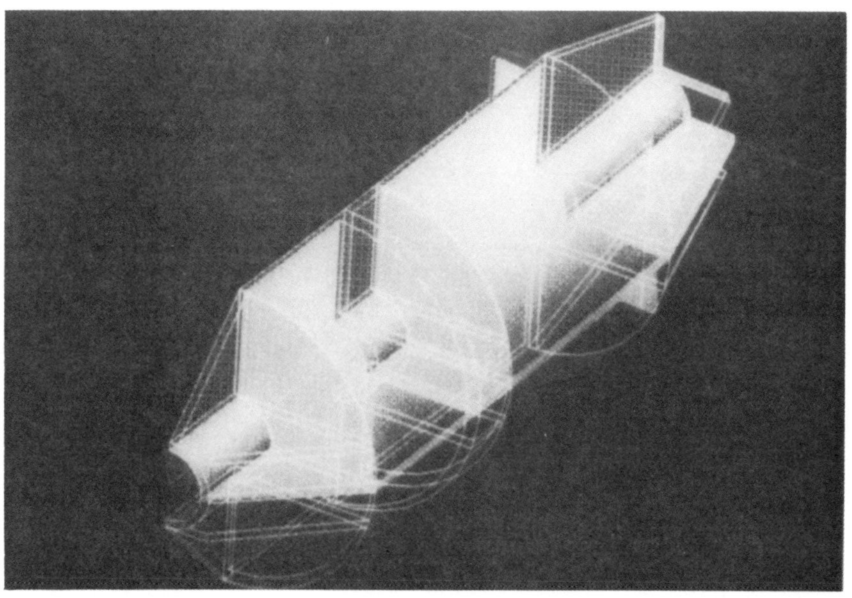

Figure 4.   Rotary press electric motor solid modelling display.

PRACTICAL EXAMPLES OF CADCAM IN MOULD AND DIE DESIGN AND MANUFACTURE

G.J. Brownhill*

This paper has been orientated towards practical examples of Computer Aided Design and manufacture in Mould and Die design - manufacture.

To achieve this I will be explaining what tools are available in CADCAM to help this industry, whilst at the same time illustrating these principles with practical examples from McDonnell Douglas's UNIGRAPHICS CADCAM system. I believe this will achieve two things. Firstly, a greater understanding of what CADCAM has to offer; secondly, an appreciation of how this Technology is applied to this area of industry.

INTRODUCTION

To achieve these objectives I have separated the Mould and Die design/ manufacturing process into 4 areas.

1. Component Design
2. Analysis
3. Mould Design
4. N. C. Graphic Machining.

This does not represent a flow path of the design process but distinct areas in which CADCAM has products to offer. The first three areas tend to be highly interdependent.

Most mould and Die engineers would use CAD/CAM in all of these areas. They may well find that their present methods of operation will change to take best advantage of the opportunities that CAD/CAM will present. Typical of those changes are that the designer will become more involved in the piece part design process, or that greater use is made of 3 axis and/or even 5 axis machine tools in the manufacture of the Die.

To understand these changes and their implications more fully, let's start talking in detail about the 4 stages of Design and Manufacture.

COMPONENT DESIGN

Most moulders are in fact a service industry, they provide a service typically to people such as the motor and telecommunications industry. Therefore, they

---

*McDonnell Douglas Information Systems Limited

will tend to do very little original piece part design. The industries they
supply have, in the main, already installed a CAD or CAD/CAM system for
designing the piece part. It therefore is logical to take this data from
the customer and feed it directly into the mould designers CAD/CAM system.
The industries they supply unfortunately do not all have the same type of
systems. In fact some companies even have different types of systems within
their own operations. The data must therefore be converted into a neutral
format to allow it to be read into the mould makers CAD/CAM system. The most
widely used format is IGES (Initial Graphic Exchange Specification).

IGES in its early stages developed a poor reputation for its quality of
data exchange. But I believe it is fair to say that, whilst IGES is not
perfect it has come of age and is now exchanging geometric data reliably in
the field. Whilst IGES is the most commonly used format, new formats are
being developed and released to widen the scope of neutral formats. Indeed
IGES is constantly being enhanced to widen its scope and capabilities.

As I indicated earlier these neutral formats are not perfect. One of the
major reasons for this is that these formats are designed to try and cope
with all CADCAM system suppliers (well over 100 at the last count) and
therefore the neutral formats like IGES cannot cope with 100% of all cases.
For reasons such as this custom interfaces have been developed that can provide
two way interfaces specifically between CAD system A and CAD system B, thus
providing a more reliable and extensive interface.

As a result of these types of interfaces, neutral formats and custom
interfaces, mould makers can have a dialogue with their customers by
transmitting data to and from their respective CAD systems. This data will
usually come in the form of a 3D wireframe model. The mould maker can
interrogate this model, add his required modifications, such as draft angles,
and then pass this data back to the customer. The customer can then examine
the revised 3D model and check to see if these changes are acceptable. This
dialogue can take place at any time and as many times as necessary.

This data would normally be transmitted via standard ½" magnetic tape but
this can be taken one stage further and, data transferred, via a telephone
link between customer and supplier, thus forging stronger customer to
supplier links.

If data were not available from a customer's CAD system then typically design
data would be given to the moulder on a conventional 2 dimensional drawing.
These drawings would then be input into the mould designers CAD/CAM system
by utilising construction techniques not too dissimilar to those used to create
a normal 2 dimensional drawing.

Finally, on the subject of component design, I started this section by
stating that mould and die designers do very little, if any, original piece
part design. the experience with users of UNIGRAPHICS CADCAM system in this
industry has shown that this is an area that does change. Initially because
of the closer links forged from the ability to exchange 3 dimensional data,
moulders have found that they get involved much earlier in the design cycle
with their customers and thus, have an increased influence on piece part
design.

Also many users have found themselves involved in the actual design of
piece parts. Manufacturers of these components have approached mould and
die makers with an urgent request to produce moulds for these components.
They may have presented the moulder with rough sketches of the component

and asked them to provide the finished article. CAD/CAM allows the moulder
to react very favourably to these requests. This is due to the increased
productivity of CAD/CAM and hence lead times from Design to Manufacture are
dramatically reduced.

ANALYSIS

Analysis in CAD/CAM can mean many things, from finite element modelling to
kinematic analysis of mechanisms. However, analysis when applied to mould
making usually means the analysis of plastic flow in an injection mould.
Benefits that have been identified from experienced users of analysis
programmes are:- better part quality, reduced tooling lead times, optimum
cycle times and reduced rework and scrapping of tools.

   There are a number of programs available for predicting plastic flow in
injection moulds, a few examples are:-

1.   MOLDFLOW from Moldflow Australia Pty Limited.
2.   FLOW ANALYSIS from Plastics & Computer Inc.
3.   SIMUFLOW from Graftek Inc.
4.   MOLDFILL from Application Engineering Corp.

   Moldflow is probably the most widely known and used programme. These
programmes can be run on a simple IBM PC, but at this level they provide a
limited graphical capability in terms of creating a model and then displaying
the results. Moldflow is fully integrated into McDonnell Douglas's UNIGRAPHICS
CAD/CAM system and in this guise overcomes the graphical limitations. Other
benefits gained using analysis programmes integrated with CADCAM are that
the analysis would be performed on the 3D model created in the component design
process previously described. Another advantage gained would be the time
required to process an analysis. Because CADCAM systems typically run on
mini-computers and not personal computers, analysis can be processed in minutes
rather than hours, thus allowing far more options to be evaluated in any given
time.

   Analysis of a model can take on various levels of complexity. A very simple
analysis would be where the parameters of the maximum flow length in the cavity,
a single thickness for that flow, mould and melt temperature would be specified.
Prior to this the material to be used would have been specified. This material
would have been selected from a library of materials stored on the system.
For this type of analysis the systems computes a series of injection times
and for each time generates the following:-

1.   Pressure required to fill.
2.   Shear stress of the plastic at the start of fill.
3.   Shear stress of the plastic at the end of fill.
4.   Melt temperature at the end of fill.

   By examining the results one can define a range of injection times within
which to operate. If, from the initial set of results obtained, an appropriate
set, or range of conditions cannot be found, the analysis can easily be run
again with a change of material, melt temperature or mould temperature or,
a combination of any of these parameters.

   This would be simplest form of mould analysis; but there are many types
with differing levels of complexity of analysis. A complex level of analysis
would be a full 3 dimensional finite element analysis.   From the 3 dimensional

wire frame model that has been already created on your CADCAM system, a finite
element mesh is generated. A full analysis of the model can then take place
by the user specifying the following:-

1.   Material to be used.
2.   Injection point or points (multiple injection points are allowed).
3.   Injection time.
4.   Melt temperature.
5.   Mould temperature.

 The results obtained from this analysis are:-

1.   Pressure drop in the mould.
2.   Melt temperature within the mould.
3.   Position of the material flow plotted against time.
4.   Shear rate.
5.   Shear stress.

 From the analysis results the mould and die designer can identify a number
of features. these results can either be tabulated or more usefully displayed
graphically on the 3D model as contours. Each line of contour would represent
a constant line of pressure, temperature, flow, shear rate or shear stress,
depending on the results the user elects to see. Each type of result
indicates to the moulder different aspects as to how the plastic has behaved
within the mould.

 Absolute pressure and pressure drop can indicate whether the mould can be
filled and highlight where die flashing may occur. It can also show overpacking
of one part of the die which results in differential shrinkage and localised
stresses that can lead to warpage within the finished part. Melt temperature
can be used to identify localised hot spots. It can also show where the flow
of plastic goes below , or dangerously near, the no-flow temperature of the
material, thus preventing short shots occurring.

 The advantages of using mould analysis programmes in conjunction with CADCAM
are numerous. The most obvious advantage comes from the ability to simulate
and check the various options available before cutting metal for the tool.
It can also provide the mould and die designer with a powerful tool in the
dialogue with the customer. Instead of arguing that "I believe that the design
is unmouldable under certain conditions", it can be demonstrated analytically
that the mould is unmouldable. The designer then demonstrates to the customer
that with certain modifications, maybe the addition of a gate, the mould can
become mouldable. This level of technology and service can only improve the
customer's confidence and relationship with suppliers.

<u>MOULD DESIGN</u>

Once the geometric data has been defined in the CAD/CAM system the mould can be
designed around it. Mould design is an area where CAD/CAM can be applied
to a very productive level. From the 3D model 2 dimensional views can be
derived easily. UNIGRAPHICS II automatically creates the six standard
orthogonal views plus two isometric views. The user can also define a limitless
number of extra views of the model by simply defining the viewing plane required.
Sectional views can also be easily derived from a 3D model. The user indicates
geometry to be sectioned, then by defining the plane by which the geometry
would be cut, can automatically create a sectional view.

Mould plate design using standard plate components is a perfect application
for parametric programming. GRIP is the parametric programming module available
within UNIGRAPHICS. This provides the user with the ability to automate the
functions available in UNIGRAPHICS. One of the primary features of GRIP is
its ability to allow the user to interact with the programme as it is running.

A typical application of GRIP would be capturing a design parametrically
and thus enabling an infinite number of variations to be created easily and
quickly.

This parametric programming facility saves an enormous amount of design/
draughting time. At the same time, all design criteria can be met and error
checks performed automatically. One such application of GRIP in the area
of mould design has been the capture of standard plates, pin, bushes etc.,
from standard suppliers such as HASCO, DME and Uddeholm.

Because of the interactive nature of GRIP these programmes are very simple
to run. Firstly, the user would be asked for the plate size to be used, which
he can determine from the 3D model. If a non-standard size has been input
the system could indicate the nearest sizes available or list the available
sizes. The programme would then take the user through a series of question
and answer sessions asking for information on wideths, plate type, steel grade,
locations and whether a section and plan view would be required. At the end
of this a plate assembly will automatically be created in as many 2D views
as required by the user. Guide pins, ejector pins, guide bushes and water
lines are also added to complete the design. During the mould assembly creation
a Bill of Materials will also be created and available for reference at any
time.

Thus in these two areas - view creation and mould plate design - significant
time saving can be achieved over existing methods.

N. C. GRAPHIC MACHINING

NC programming has always been acknowledged as being the area where it has
been straightforward to see the benefits from the installation of a CADCAM
system. The NC programmer starts where the designer leaves off. The programmer
generates the NC data by graphically driving a tool over the 2 dimensional
or 3 dimensional model the designer has already created.

Within UNIGRAPHICS there are two software modules available for aiding the
programmer in creating the cutter paths. These modules are GMM, for 2, $2\frac{1}{2}$ axis
machining and GMAX for 3 through to 5 axis machining.

The Graphic Mill Module (GMM) gives the user the ability to perform, point
to point operations, profiling, pocketing and ZIG/ZAG machining in a 2 or $2\frac{1}{2}$
axis motion.   The user would be asked to input such features as clearance
planes and engage and retract vectors.   All of these operations can be
performed on 2 dimensional geometry with the programmer adding 3 dimensional
data (such as clearance planes) as and when needed.

The Graphics Multi Axis Machining (GMAX) provides the user with a full
3 dimensional machining capability.   This allows tool paths to be generated
on any Unigraphics surface.   The tool axis can be defined as being square
to the bed of the machine tool (3 axis), as an arbitrary fixed vector (4 axis)
or normal to the part surface.

Many options are available in GMAX.   Surface contouring provides a method
of generating a tool path over any single surface.   The tool centre line
will be controled such that it follows a user specified path in such a manner
that the tool will never gouge the surface.

Parameter line machining provides the ability to machine multiple surfaces.
By using this method, the user has two options for controlling surface finish.
 The first being by specifying the number of steps in the cut and step
directions, whilst generating the path.   Secondly, the user can specify the
maximum deviation the cutter can make from the surface in the cut direction
and control the stepover by the height of the scallop to be left by the cutter.

For all of these modules, a man readable code will be automatically generated
in the APT programming language format by UNIGRAPHICS.   This code has to
be converted into machine readable code via a post-processor.   A post-processor
will be required for each different machine-tool/controller combination on
the shop-floor.   Most CAD/CAM vendors have a library of "off the shelf" post-
processors (UNIGRAPHICS has a library of over 2000) and may be able to supply
immediately with a proven post.   Experience has shown that typically, a new
post-processor would almost certainly have to be created for one or two machine
tool.

The problem with this approach has been the time element.   Typically, it
would take three months to create a new post-processor.   Therefore, there
has been a trend for vendors to develop Generic Post Processors.   Whereas
most require the user to write a programme in some sort of code, maybe FORTRAN,
McDONNELL DOUGLAS has used its experience gained from writing over 2,000 post-
processors, to develop an easy-to-use Generic Post Processor.   Using this
module, the user can quickly create a post-processor, depending on the
complexity of the machine tool/controller combination.

The user creates the post-processor by means of a question and answer session.
The module will ask the user such questions as, which axes are valid, what
G and M codes are valid for which operation, and even allow the user to input
special commands.   This gives the Mould and Die manufacturer complete
flexibility and control over the post-processors.

The standard method of putting post-processed information into machine tools
has been via paper tape.   This has limitations.   To overcome these limitations
DNC links (Distributed Numerical Control) have been developed.   DNC offers
many advantages to the machine shop, by being able to take post-processed
data from the CAD/CAM system and feed this directly into the correct machine
tool controller.

A DNC link, such as Masterlink 3000, can control up to 16 machine tools.
DNC eliminates the need for paper tape altogether, but that will not be the
only advantage gained.   A DNC system is a powerful micro-computer and will
have extensive editing facilities available to the user.   Across the 16
machine tools controlled by the DNC link, up to 16 different controllers with
16 different editing capabilities will possibly be available.   DNC will reduce
this to one common powerful editing capability across all the machine tool
range.   Because the DNC and CAD/CAM system can give all the features to be
found on the best controllers, there will be no need to duplicate these features
on the machine tool.   As a result, machine tools can be equipped with far
less intelligent controllers whilst still retaining all the capabilities.

DNC can store the equivalent of many thousands of feet of tape locally.
The data can be down loaded on an "as and when required" basis to any machine
tool, thus overcoming any memory limitations your controllers may have.
DNC is the final link to total flexibility, control and capability in CAD/CAM
in Mould and Die manufacture.

CONCLUSION

Design to Manufacture using one database.   That is what CAD/CAM can give
you.   The design can be refined or even created within your CAD system.
Analysis to predict the flow of plastic can be performed on the same 3D model.
The mould tool can then be built up around this model.   Finally, NC machine
tools can then be driven and controlled by data generated from the CAD/CAM
system.   This capability allows the Mould and Die design manufacturer to
maintain a high quality and quick response time and provide the customer with
a very much higher level of service.

Author's biographical notes

1975-79   4-year mechanical apprenticeship with Jaguar Cars.
                 At the same time, gained an HNC in Mechanical Engineering.

1979-84   Responsible for concept design on XJ40 chassis and interior
                 trim components.   Much of the design work was carried
                 out using the CAD system, installed in 1981.

1984-present   Joined McDonnell Douglas as a CAD/CAM consultant.   The
                 last year has been spent specialising in Computer Aided
                 Engineering, with special interest in the plastics industry.

INTERFACING—THE CHALLENGE FOR THE CAD/CAM INDUSTRY

Michael Abrahams* and Jim Macintosh*

The paper describes the need for interfaces and comments in detail on the use of IGES, VDA and custom interfaces for the direct transfer of geometric data which will be essential if the benefits of CAD/CAM are to be enjoyed by the Plastics Industry. A number of practical examples, selected from automotive and plastics components are presented in detail.

1. INTRODUCTION

For a number of decades the polymer industry as a whole has enjoyed significant growth which has been the envy of traditional materials; this period is now over, and it has become necessary to improve efficiency, reduce costs and improve productivity in all segments of the industry from polymer manufacture to polymer processing. The results of these changes are clearly evident from the closure of inefficient chemical plants to the use of robots at the moulding press but there is still a long way to go before the industry provides the returns on investment required by its financial backers.

Today no-one can argue against the increase in efficiency that CAD/CAM systems have brought to the design, engineering and manufacturing process. Benefits result at every stage from lead-time reduction, fewer prototypes, better quality, improved reliability and reduced costs.

The mechanical engineering industry has been investing in CAD over the last five years and today the Plastics Industry is beginning to take the first investment steps being driven by their end-user customers. In the auto industry the drive to replace steel by lighter weight, more cost effective materials, is growing and thereby increasing the demand for plastics. The auto companies are now demanding that their suppliers, i.e. the plastics moulders and tool makers, invest in design and manufacturing technologies, CAD/CAM, to become more efficient and to take user design data directly on magnetic tape.

* General Electric Calma

However, just investing in any "CAD/CAM system" is not the simple answer.
Most designers produce a precise definition of specific parts in house, and
then contract out the tooling and the production to the toolmaker and the
moulder respectively. If the designer and his supplier have the same
vendor's CAD/CAM system, exchange of precise geometric data is easy.

Nearly all toolmakers and moulders provide parts to more than one end-user
and it is highly unlikely that <u>all</u> the end-users will have the same CAD/CAM
system as his suppliers.

So the supplier has a major problem! He could install systems to match
those of each of his clients - but if he were so impractical, he wouldn't
continue to be in business because of duplication of investment.   The
obvious solution is to invest in a CAD/CAM system which not only meets his
design and manufacturing needs but is also able to interface with all other
major vendor or in-house systems. By communicating electronically with any
other system the supplier avoids all the problems of acquiring and working
from designers' drawings.

2. <u>WHY</u> <u>A</u> <u>GOOD</u> <u>INTERFACE</u> <u>IS</u> <u>IMPORTANT</u>

Interfacing is a highly complex process. Different vendors' CAD/CAM systems
use different data formats, obey different protocols and have different
philosophies for representing geometry and in many cases use different
computers - and these are just initial hurdles! The interfacing process can
often introduce errors into a part definition.   Such errors could be
reconstructed in the supplier's CAD/CAM system and prevent him from
complying with the designers specification.

It is essential that the sorts of problems likely to occur are fully
understood. Both suppliers and designers can then take full advantage of
all that interfacing offers over traditional methods of exchanging drawings.

One advantage for working for one of the world's largest diversified
manufacturing companies is the experience gained on all the major CAD/CAM
systems available in the world today. General Electric have had to amass
considerable expertise in the development and the use of interfacing for
different vendor systems. For example, General Electric helped develop the
I.G.E.S. (Initial Graphics Exchange Standard) specification and continues
to be active in the continuing development of this standard.   GE Calma has
therefore been able to draw on the GE experience and now offers interfacing
tools which will fully meet the needs of the Plastics Industry.

Calma's Interfacing Tools

General Electric Calma's interfacing tools fall into three main categories:

* IGES (Initial Graphics Exchange Specification) - the CAD/CAM
  industry standard;

* VDA (Verband der Automobilindustrie e.v.) - the German automotive industry standard;

* Custom - special interfaces for the direct transfer of geometric data.

3. <u>IGES</u>

IGES is a neutral data format for describing product design and manufacturing information. The IGES format is designed to be independent of all CAD/CAM systems, so that information created and stored in one can be read and used by any other CAD/CAM system.

To avail of the IGES format, users must install translators that link the dissimilar systems. One translator, called an IGES preprocessor, translates data from the sending system's geometric data base into IGES format. Another translator, called an IGES postprocessor, translates the IGES data into the receiving system's data base format.

Some CAD/CAM vendors offer both pre- and/or postprocessor software. Other vendors do not supply any translation software and users therefore have to develop processors in house if they wish to implement IGES. This can be a time consuming and expensive process which is prone to error and should not be contemplated by the typically medium sized companies in the Plastics Industry.

The first version of IGES 1.0 appeared in 1980, and was adopted as part of a USA standard (Y14.26M) in September 1981.

The basic elements of data being transmitted using IGES are called entities. An Entity can be as simple as a description of a line, or as complex as a description of a complete sub-assembly.

The IGES specification covers geometry, anotation, i.e. text, and structure, i.e. position. Initially the only geometric entities which could be transferred were simple line drawings and there was no capability to transfer 3D models.

In January 1986 the specification for IGES 3.0 was published, this specification refines and extends the entity definition missing from the two previous releases. In particular, full support is now being offered for surface transfer in full 3D, improved anotation and complete finite element definition. General Electric Calma will be implementing this latest version of IGES 3.0 in their next major software release this year. For those plastics companies who currently use IGES this new specification particularly in the areas of surface generation and N/C tool path creation, are at last addressing manufacturing specification problems which have hindered its previous credibility.

<u>Limitations</u>

The IGES programme is very large and highly ambitious. No programme with the size and scope of IGES is going to be trouble free. IGES is no exception. Its present shortcomings can most easily be summarized under the following headings.

* File size and complexity: IGES files are often large and very complex. Thus IGES processors are rarely easy to write, while handling IGES data can be cumbersome. Moreover, the results are only as good as the worst of the IGES processors available to the user.

* Entity support: This problem is caused by the fact that IGES is not a master set containing every entity that all CAD vendors use. Rather it is a general recommendation for a neutral data format. It is therefore open to differences in interpretation, both from vendor to vendor and from user to user. These can sometimes result in differences of opinion about how data should translate.

An obvious question is, "If IGES has these limitations, why use it at all?" The reasons are plain. IGES is in use by an increasing number of design and manufacturing organisations. Indeed, more and more large firms are demanding IGES compliance from CAD/CAM vendors. These companies realise the crucial role IGES will play in their design and manufacturing capability which will enhance productivity, accuracy and hence quality.

4. VDA/VDAFS

Background

The IGES specification suffers from its attempts to be universally applicable. Certain industries rely heavily on accurate computer representation of their products, and IGES today is incapable of addressing the specific requirements of all of them.

The German Automotive Industry is a case in point. Realising that IGES was not meeting their advanced surfacing and numerical control (NC) needs, a group of German car manufacturers formed a working group in 1982. Their aim was to develop a manufacturing standard which would unify, not only the automotive industry, but also its supply base which includes the toolmakers and moulders.

The working group included representatives from Volkswagen, Porsche, BMW, Opel, Audi, TH Darmstadt, Daimler-Benz and WMI. These firms were already mature users of CAD/CAM systems, and so had the knowledge and experience to achieve their aims.

What finally evolved from their efforts is known as the VDA (Verband der Automobilindustrie) standard. Such is its success that other European companies have also decided to adopt it. Now VDA looks very much like becoming an international as well as a national standard.

Special Features

VDA is a manufacturing standard, specifically oriented to the automotive industries. The standard is essentially very simple and supports only a limited number of geometric entities all of which are described by polynomials.

In general, surface geometries can be defined in one of two ways: explicitly or procedurally. If the geometry is defined explicitly, then the information stored is largely numeric and is described by some underlying polynomial, e.g. Bezier surfaces.

Procedurally defined geometry relies less on numeric data, and often makes use of other geometry which has been previously defined, e.g. ruled surface.

CAD systems which contain procedural geometries must use uniform polynomial approximation techniques to transform their geometries into the VDA format.

The VDA specification ensures that the integrity of the geometry being transferred is not compromised, and tool path generation can proceed with the minimum of problems. Explicitly defined geometries, in particular, are transferred error free.

If polynomial approximation is required, then the level of approximation must be under user control so that he can regulate the accuracy of geometric transfer. This is of course fundamental and very few CAD vendors offer this facility.

5. CUSTOM INTERFACES

Before the creation of any standard interfacing procedures, users themselves had to answer the question of how to transfer data from one incompatible system to another. Even today, the available standards fall short of some users' requirements in certain areas.

Nor is the problem simply one of communication between competing vendors' systems. Many users need to swap data between different applications as well.

Satisfactory data exchange depends on two basic factors:

* the systems must have access to a database of information stored in a neutral format;

* the information must be classified in a logical way for ease of selection.

A first approach was to work around the question of geometric description by breaking down a given structure or material into a finite set of discrete elements. This approach forms the basis of the Finite Element Analysis method.

Appearing at the end of the 1950's the method represents each surface by a mesh of elements, linked by a number of control points. The relationship between these points and the original structure must be very well defined, so that the meshed model itself will reproduce the phenomenon being studied by finite element techniques.

Once the finite element mesh has been created, it is straightforward to format the data that describes the structure under investigation. The data appears as preprocessor output and consists mainly of points (i.e. three-coordinate sets) and connectivity information (e.g. point relations, lines, curves and surface definitions).

For some well-defined CAD applications, the finite element mesh has turned out to be the most efficient technique for data exchange and is still retained for mesh transfer for structural analysis purposes.

For more complex geometry descriptions, however, the finite element method is not applicable.

A review of the European automotive industries will reveal that the majority of the CAD systems which are in use for sculptured surface design have geometric data bases which are defined purely in terms of polynomials. The transfer of data from such a system is straightforward requiring only the extraction of the polynomial coefficients.   This extracted geometry is then rebuilt exactly in the receiving system provided that system has a compatible polynomial descriptor data base, e.g. from Renault to Volvo whose in house systems are based on Bezier mathematics.

For those CAD vendors and their customers whose geometric data bases do not support the same polynomial description as that of the sending CAD system the problem of data transfer is difficult.

These problems do not exist in a General Electric Calma system because the user has the unique capability of defining the polynomial description which exactly matches that of the sending CAD system.

For those CAD systems whose geometric data bases are a combination of both procedural and polynomial geometries, direct conversion requires geometries to be "mapped" before they can be transferred. Such are the complexities of this approach that the development of convertors now tends to be handled by third party specialist software houses. As well as being able to receive data from a different system, it is important that the receiving system can also send data back to the original system in a similar format.

6. <u>EXAMPLES</u> <u>OF</u> <u>INTERFACING</u>

1. IGES transfer - Ford PDGS to G E Calma DDM
   Moulded Xenoy Bumper bar - Ford Sierra

2. IGES transfer - Ford PDGS to G E Calma DDM
   Windshield (1/2 model)

3. VDA transfer - Volvo Catia to G E Calma DDM
   Volvo 740 fully-surfaced front hood steel pressing

4. Direct transfer - ARG neutral file to G E Calma DDM
   Complex surface test model

5. Direct transfer - Renault Unisurf to G E Calma DDM
   Fully surfaced side fender

6. Direct transfer - Renault Unisurf to G E Calma DDM
   Fully surfaced side fender - detailing tool path generation

7. SUMMARY

This paper describes the background and need for interfaces between competitive CAD/CAM systems and is based on the authors combined experiences in the Automotive, Plastics and CAD/CAM Industries. Data transfer is the essential first step in a process which must ultimately result in the manufacture of the designed part. A good interfacing capability which then uses a common design and manufacturing data base must also be a prerequisite for any CAD/CAM system suitable for tool makers and moulders in the Plastics Industry. Those companies who now are investing in CAD/CAM for the first time must carefully evaluate the interfacing capabilities of their prepared CAD/CAM vendors. It is likely that they will have a number of end-user clients who will each have different CAD/CAM systems thus good interfacing capability in their chosen system is essential to maximise the return on their investment.

As well as a good interfacing capability the system must also be able to satisfy the particular needs of the industry in the areas of mould design, mould cooling and polymer flow analysis and tool manufacture using multi-axis machine tools. In the authors experience there are very few of the 300 plus CAD vendors in the market place which fully meet the needs of the plastics industry. As the price of fully integrated CAD/CAM systems decreases, the Plastics Industry should invest now if it is to maintain a technological edge but only after evaluating the few systems which <u>totally</u> meet the unique requirements of this industry.

<u>Acknowledgement</u> The authors wish to acknowledge the help and assistance provided by Emmanuel Pavageau and his projects team who developed the VDA and direct interfaces described in this paper.

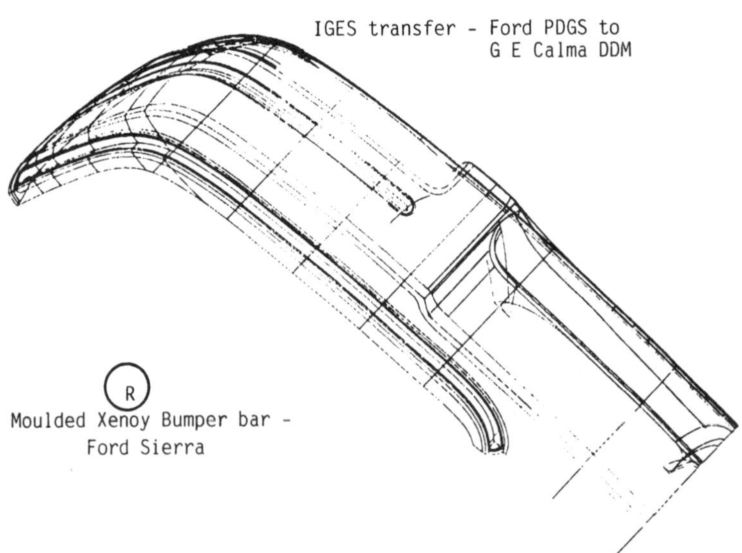

Figure 1

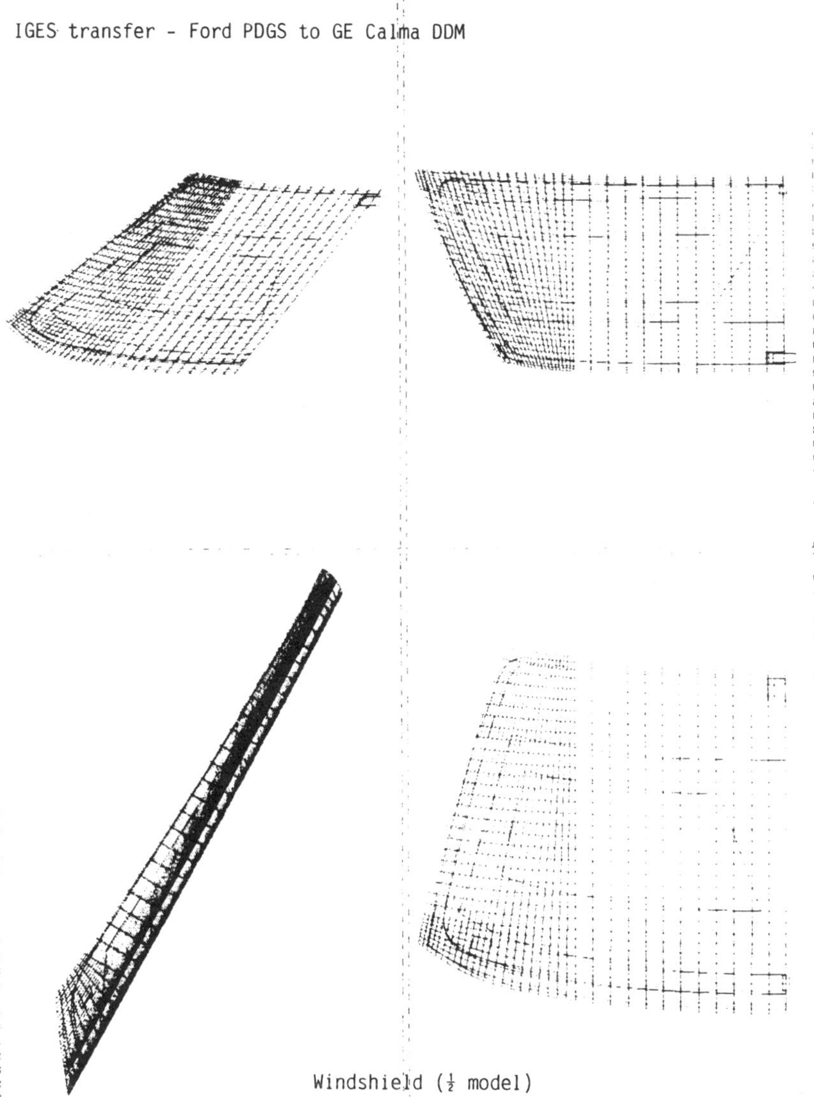

Figure 2

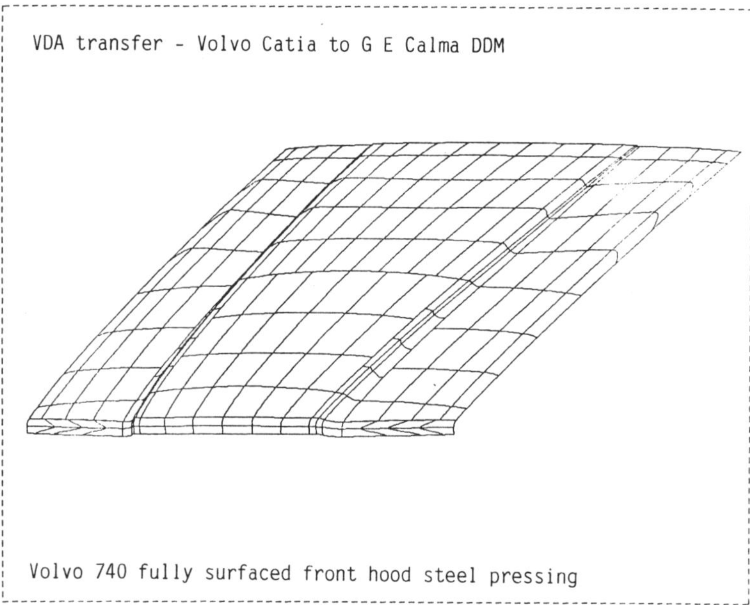

Figure 3

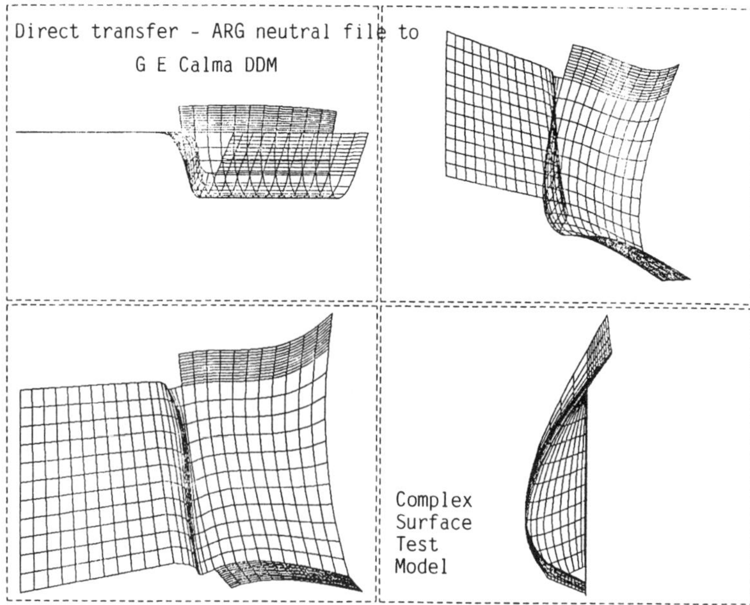

Figure 4

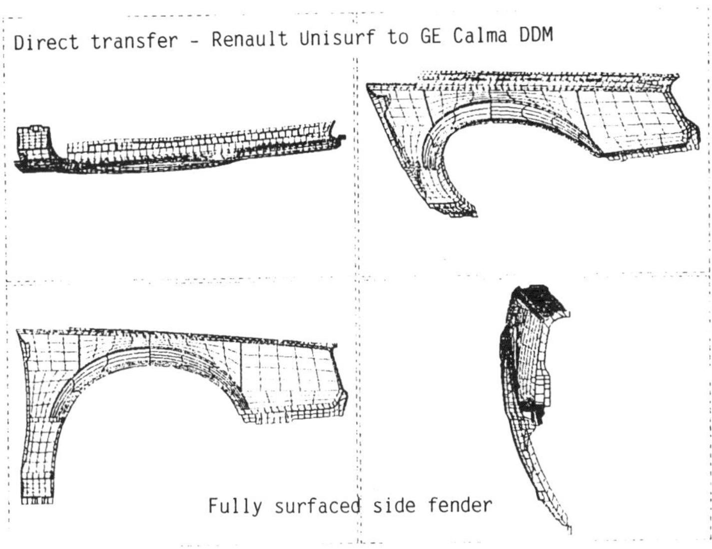

Figure 5

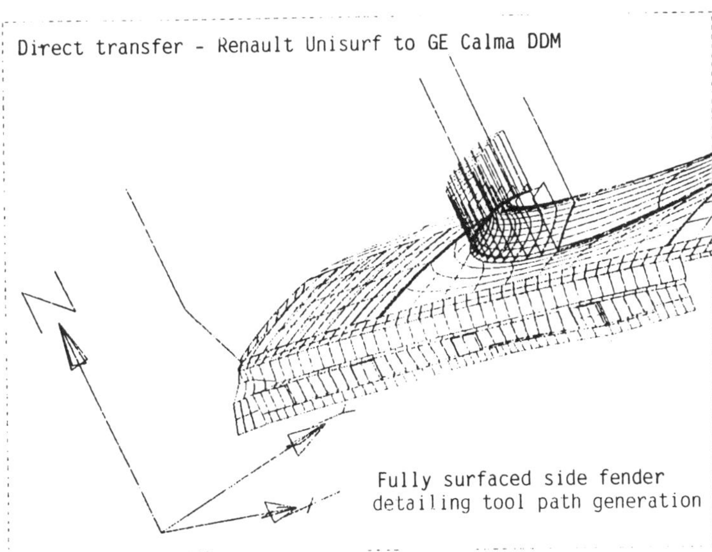

Figure 6

COMPUTER SIMULATION OF INJECTION MOULDING

I.T. Barrie*

A computer program** is described which helps
moulders answer the main questions :-

    Is the mould design feasible and optimum?
    What are the optimum moulding conditions?
    What are the minimum machine requirements?
    What is the likely production rate?
    What would be the likely manufactured cost?

With a complete quantitative simulation of a
moulding trial by computer, many important
decisions can be made with the design still at
the drawing board stage, before metal is cut.

<u>INTRODUCTION</u>

The injection moulding industry has evolved to its present state
of technological development relying heavily on the established
skills of toolmakers and machinery makers for the metal-ware, the
new devices of the electronics industry for process control, and
on the advice and technical support of the plastics materials
suppliers for product development and design. Yet after over 40
years there is hardly one moulder in the world who understands
and uses the information available on plastics material flow
behaviour and thermal properties to minimise risks and errors
regarding product design, mould design, production planning or
costing.

The reasons why the known and published material properties have
been ignored are manifold, but one major one is the complexity of
calculations that must be done to make theory and empirical laws
yield results which are better than the 'rules-of-thumb' evolved
over the years by experience and trial-and-error.

The modern 'business' micro-computer systems are now sufficiently
powerful, quick and cheap to do the calculations. This paper is
concerned with using a computer for the benefit of the injection
moulding industry by means of a fully interactive, user-friendly
system of programs which allow a complete quantitative simulation
of the moulding process, ending up with a product costing.

*   Ian Barrie Consultancy
** 'SIMPOL' Injection Moulding Simulation System

MAIN OBJECTIVES FOR SIMULATION OF THE MOULDING PROCESS

The main objectives for computer simulation are taken as those
arising in practice in a mould-proving trial before commissioning
for full production. These are summarised as :-

    (a)   Is the product designed for optimum performance <u>and</u>
         ease of production?
    (b)   Does the mould design, configuration, gating ensure
         feasibility of mould filling for the chosen material?
    (c)   What are the moulding conditions for optimum cycle
         times and cost?
    (d)   What are the minimum machine requirements?
    (e)   What is the likely production rate?
    (f)   What is the breakdown of fabrication costs, material
         costs, regrind rates?

The `SIMPOL' program provides quantitative answers to these
questions based on material flow, thermal and density data (from
raw material suppliers or testing house), cavity geometry (user
supplied from 2-dimensional layflat equivalent to moulding
geometry) and moulding conditions (user selected with computer
guidance). Notes are given in the Appendix on the generation of
data files for material properties, cavity geometry and machines.

PROGRAM STRUCTURE FOR ANALYSIS OF THE PROCESS

Figure 1 shows the logical sequence for the input of information,
analysis of the data, interactive choice of moulding conditions,
and output of results leading finally to the estimation of
production costs.

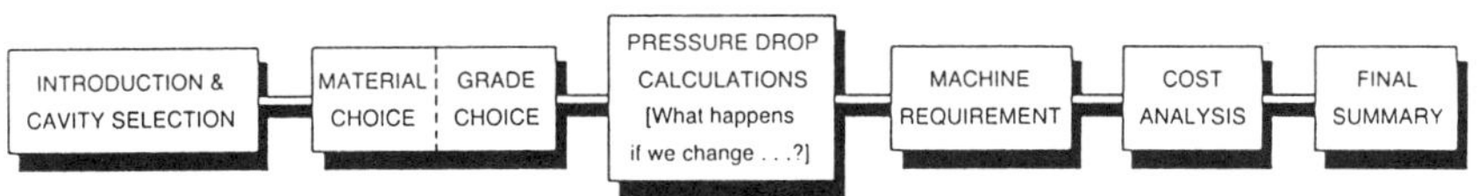

<u>FIGURE 1</u> Program structure for simulation of the moulding process

Data are picked up as required from the pre-prepared data files
relating to mould, material and machinery data. Material data
remain available for all cavity geometries to be called upon as
one would use a bag of polymer from the material store. Likewise
a machine specification on file remains until that machine is
scrapped or replaced. A cavity file is prepared for each
moulding, including hot-runner and/or cold-runner dimensions.
The cavity thickness and the runner sizes (not lengths) can be
altered interactively within the main simulation program. Also a
cavity file itself can be quickly edited to reflect design, gate
or runner changes.

THE IMPORTANCE OF GOOD MATERIAL PROPERTIES DATA

The injection moulding process, like all other thermoplastics
processes, comprises three stages – heating, shaping, cooling.
It is self-evident that any serious analysis or synthesis (ie
simulation) of the process in quantitative terms relies heavily
on accurate knowledge of the material flow behaviour and thermal
properties. It is fortunate that good data is usually available
from the main materials suppliers, trade associations and
academic establishments; if you put bad data into a calculation
you can only get bad results out.

The calculations using these and other relevant data are
particularly complex for injection moulding because of the
interaction between the various quantities. The following are
some examples illustrating the need for a thorough understanding
of what is happening in order to begin to calculate the
quantitative effects.

Melt Preparation   Heat is supplied to raw polymer chip or powder
in two ways – by conduction from the walls of the heated barrel,
and heat generation within the polymer due to its compression or
mechanical work (ie shear heating). Only by knowledge of both
the viscosity (hence degree of mechanical working by the screw)
and the thermal properties of thermal diffusivity and heat
requirement (called enthalpy change) can the relative
contributions of the two heat sources be assessed; and only then
can the time-limiting element be determined of either screw-
feeding rate or heat transfer rate at the metal/polymer
interface.

Simple Shear Flow   Even the flow of a viscous melt down a
uniform channel under isothermal conditions is highly complex.
With a high pressure-gradient along the channel the melt is at a
high pressure at entry and for example in the atmosphere at exit.
Thus the entering melt is compressed and has a higher viscosity
for this reason than at exit. At the same time the mechanical
work done in forcing the melt down the channel has generated heat
making the emerging melt hotter and again lower in viscosity for
this reason than at entry; this is somewhat counter-balanced by a
cooling effect on decompression. What value for viscosity should
be used in shear flow equations? How relevant are laboratory
viscosity measurements to the practical situation? Fortunately
the pressure dependence of viscosity and the associated shear
heat effects are opposite in effect and not vastly different in
magnitude, which lessens errors considerably. The rule is allow
for both effects adequately or ignore both. The greatest sin
(and error) is to allow for only one effect and ignore the other.
Flow in cold channels is a considerable further complication (see
next Section).

Convergent Flow Effects   Highly convergent flow patterns are
unavoidable, as in flow from barrel through the nozzle or through
restricted gates. A pressure loss is always associated with such
flow (the so-called die-entry pressure loss). For pin-gates for
example this component in the overall pressure requirement can be
quite large, say well over 20% of the total, and can affect the
feasibility of mould filling. No pressure drop calculation is
reliable unless both shear and convergent flow data are used.

MELT FLOW IN COLD CHANNELS AND MOULDS

A full theoretical calculation of pressures and temperatures
during melt flow in cold channels is prohibitively complex. It
is doubtful that present-day knowledge of material properties and
theory is good enough to yield worthwhile results. All practical
calculations done on modest-sized computers involve to some
degree a modelling of behaviour from empirical observations.

The 'SIMPOL' program uses a refined form of the 'frozen-layer'
model. This model was based on experimental work on instrumented
moulding machines and moulds (I T Barrie, Plastics & Polymers,
37, p463, Oct 1969) and gives good estimates of pressure drops
using measurable material thermal and flow properties. The
calculations focus on the growth of frozen material on the cold-
runner or cavity walls during filling, and the melt flow through
the resulting constricted channels (Figure 2).

The observed pressure drop in a mould as a function of the rate
of filling is illustrated in Figure 3. The pressure v rate curve
is confined by the two asymptotes given by the lines A and B.
Line A represents the limiting slow-fill rate below which melt
becomes frozen across the mould before filling is complete. Line
B represents the isothermal pressure v rate curve, ie if the
mould temperature was the same as that of the melt. In practice
at very high fill rates (very low fill times) there is little
time for heat to be lost to the walls, and the isothermal
conditions are increasingly approached. At progressively lower
rates the practical curve flattens out before climbing rapidly as
the frozen layers increase and ultimately totally block the flow.

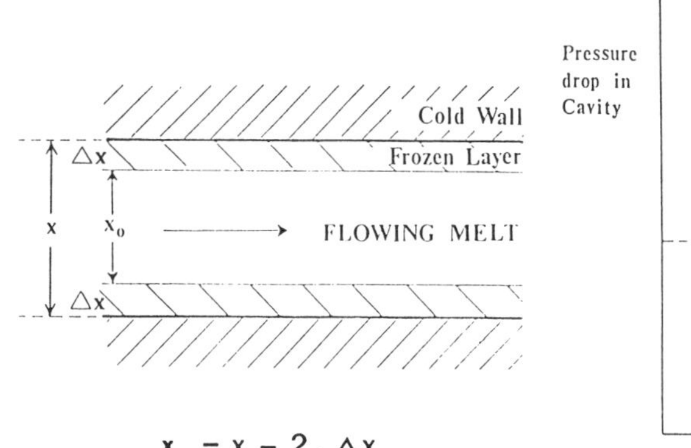

$$x_0 = x - 2 . \Delta x$$

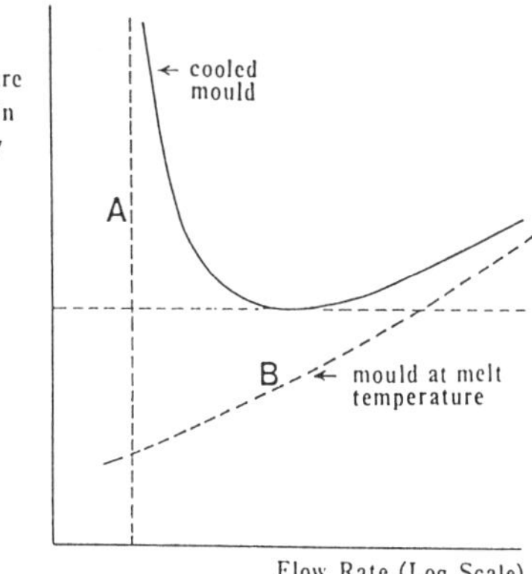

FIGURE 2 Constriction of flow
due to frozen material.

FIGURE 3 Pressure v Flow Rate
behaviour in a cold channel.

FEASIBILTY AND CHOICE OF MOULDING CONDITIONS

A knowledge of the pressure gradients in the melt during
injection is essential both for the determination of feasibility
of complete filling, and for decisions on ideal moulding
conditions. The range of sensible values for some moulding
variables is often predetermined by qualitative factors such as
surface finish, shrinkage and distortion; however within these
sensible ranges a best combination of conditions can be found, or
appropriate mould design changes may be required.

For any chosen material grade, cavity and runner geometry, and
set of moulding conditions the 'SIMPOL' program reports on the
relevant pressures, times and temperatures which result. The
maximum shear stress arising and a suitable packing (or hold-on)
pressure are also given. Figure 4 shows a printer-dump of the
screen display for medium-flow polypropylene in a car bumper
mould. [NOTE: At this stage no machine needs to be defined. In
fact the chosen conditions define the machine.]

In Figure 4 the central horizontal bar-chart shows the total
pressure-drop in the mould (hot-runner + cold-runner + cavity)
colour-coded red, blue and yellow respectively as indicated in
the schematic diagram immediately above it.

Mould Filling Limitations due to Available Pressure In
principle there is a maximum feasible pressure-drop in the mould.
If the pressure required to fill the mould is above this value, a
short moulding will result. The critical value is related to the
maximum pressure available (under dynamic conditions) minus the
pressure dropped in the injection unit delivering the melt to the
mould. This melt-delivery pressure loss is not easily calculated

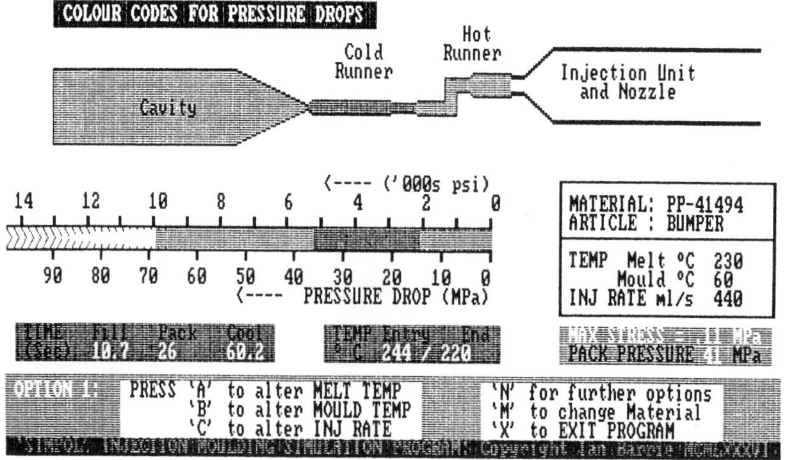

Figure 4 Pressure drops and other information on mould filling.
Actual screen display dumped to printer. (Originally in colour.)

and depends on the machine design, nozzle/check ring geometry,
the degree of wear and the hydraulics. Experimental observations
suggest that the delivery loss amounts to some 40 to 55% of the
specified maximum (hydrostatic) pressure capability of the
injection unit. Thus for a claimed maximum injection pressure of
150 MPa (1500 bar), the mould designer and moulder would not
expect any filling problems with pressure-drops below 67 MPa (670
bar) in the mould; on the other hand they would anticipate
problems with pressure drops above 90 MPa (900 bar). In Figure 4
the increasingly shaded area on the extreme left of the bar-chart
gives a visual warning of the degree of uncertainty of filling
feasibility between 67 and 90 MPa.

Quantities which can be Interactively Changed.   Many parameters
affect the results displayed in Figure 4. Many of these can be
altered interactively while running the program. They are :-

        Melt Temperature   Cavity Thickness
        Mould Temperature   Hot & Cold Runner Sizes
        Injection Rate   Maximum Available Pressure

The suggested Packing Pressure, which affects clamp force
requirements, may be over-ridden by a value of the user's own
choice. Also a 'Cooling Efficiency Factor' can be set to alter
the estimated cooling times to the user's own value. The
computer estimates cannot take into account the practical
efficiency of the cooling system actually designed for the mould;
ideal cooling channel geometry may not be possible for some mould
geometries.

An Overview of Effects of Injection Rate and Melt Temperature
An option is provided giving this overview, leading to Figure 5.

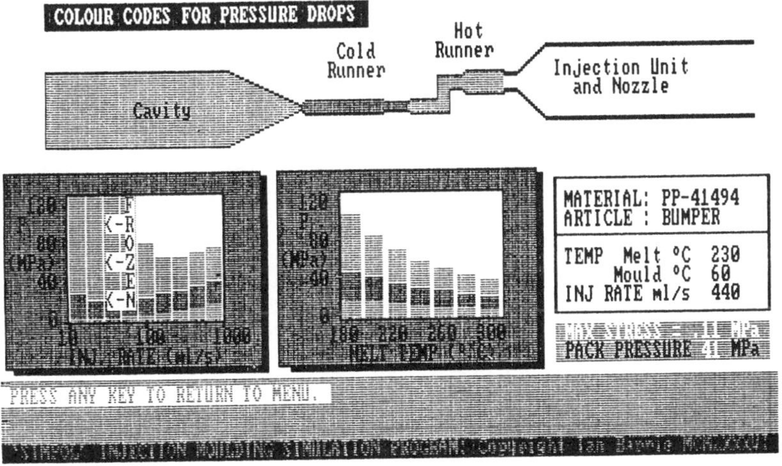

Figure 5   Bar-charts showing effects of Injection Rate &
            Temperature.

MINIMUM MACHINE SPECIFICATIONS TO SATISFY CHOSEN CONDITIONS

Minimum specifications for both the clamp unit and the injection
unit are estimated and displayed (see Figure 6). For each unit
the right-hand columns show a rounded-up typical machine size
which satisfies the requirements; this machine size is then used
to estimate fabrication costs, unless the user over-rides the
computer estimates with his own machine specification.

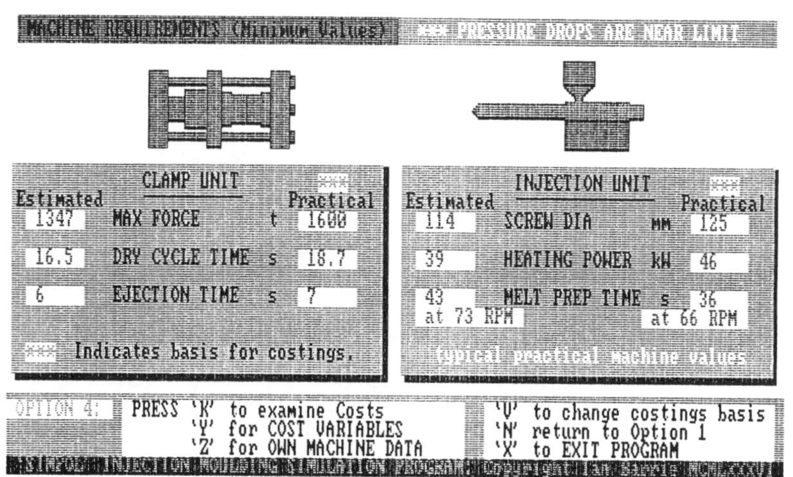

<u>Figure 6</u> Estimates of Minimum and Practical Machine Sizes.

ESTIMATION OF CYCLE TIMES

Dry-cycle time and ejection time are functions of the machine
size and the moulding size respectively. These are put with the
other times already estimated for mould filling, mould packing,
mould cooling and screw-back (ie melt preparation) to give the
overall cycle time. These values are displayed in Figure 7
(overleaf) together with cost estimates. The limiting factor,
mould cooling or melt preparation, is also indicated.

ESTIMATION OF PRODUCTION COSTS

Mould design, material choice, moulding conditions, machine
choice all contribute to the ultimate product cost via a web of
interacting factors. The 'SIMPOL' program bases the fabrication
cost on the Machine Hour Rate (MHR) concept, where each machine
earns its keep producing saleable mouldings. Realistic MHR
values are automatically calculated for cost estimates; these
values can be over-ridden by the user as appropriate, as can a
number of other cost-controlling factors.

A summary of the various cost factors is displayed in Figure 7.

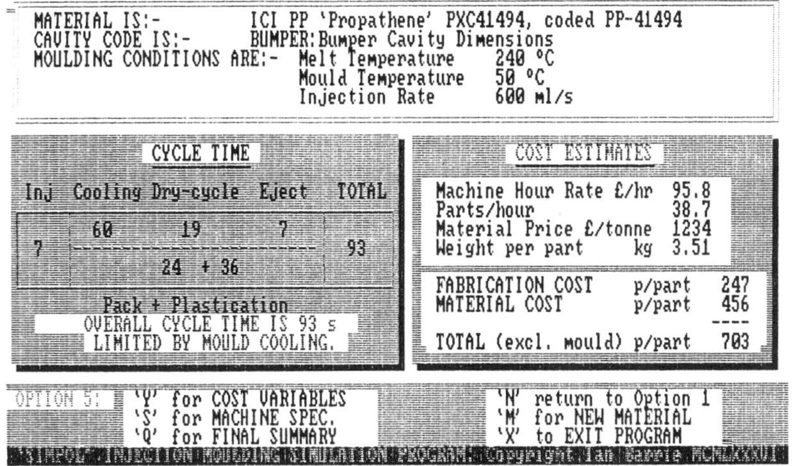

Figure 7   Summary of Moulding Conditions, Cycle Times and Costs.

A COMPUTER IS ONLY AN AID

In the context of this paper the computer is only a technical
aid, providing the injection moulder the best knowledge,
understanding and information regarding his process which
currently exists. Every means is used to achieve an non-
intrusive interface by using single-key entries wherever
possible, by never leaving the user with a demand for information
without a choice of answers, by using colour graphics and
distinctive screen layouts, by only asking questions and giving
answers which are essential and omitting all extraneous detail.

It is of paramount importance for a technical subject demanding
high expertise that requirements for computer-skills or keyboard-
skills are minimised, or even eliminated. The user must have
every opportunity to impose his own data or judgment.

COMPUTER AIDED TRAINING FOR INDUSTRY AND TECHNICAL COURSES

The complete simulation of the injection moulding process enables
a student to create his own mould geometries for his product
design and do 'moulding trials' to check feasibility and likely
costs. He will be able to link runner and gate geometry with
material behaviour under a wide range of moulding conditions. He
will become familiar with machine sizes, cycle times and costs
for various moulding sizes. In short he can gain much useful
experience through simulation during his studies which would take
a prohibitively long time and considerable expense to acquire in
a moulding shop — would be totally impossible in even the best
equiped University or Technical College.

APPENDIX - NOTES ON GENERATION OF DATA FILES

Material properties   Material flow properties specific to a
particular Grade are put on file using an ancillary program.
This program requests data for pairs of values for shear
viscosity and injection rate.   It also accepts die-entry pressure
loss data for each shear rate.   Data for up to six temperatures
can be entered for each Grade.   The 'raw' data supplied is then
used as the basis for a flow-behaviour model, the coefficients of
which are stored on file for direct access when using the
simulation program.   Poor (ie scattered) data points are queried
on entry and only accepted on the user's command.   If no die-
entry pressure data are entered, default values are used specific
to each material type.

Thermal and density data are generally not grade-specific.
Values are 'built-in' for each generic material type.

All data on file is accessible as a spot-value read-out, as a
Table as a function of temperature, or in graphical form.

Cavity Geometry files   Flow paths in the mould are stored in
two groups - runners and cavity.

Runners, hot or cold, are divided into a number of sections
(maximum 15) each of a definable geometry in terms of section and
length.   Circular, semi-circular, square and rectangular cross-
sections are accepted, with convergence or divergence allowed for
the circular and semi-circular cases.

Flow in the cavity is taken as a number of flow-steps (max 15)
each with a defined length and thickness, and with a starting and
finishing width of flow front.   This information is obtained from
a two-dimensional lay-flat equivalent model of the actual three-
dimensional part. This is an area where human brain-power can be
used to advantage; computer analysis of three-dimensional
geometries can be very expensive in software and/or very time
consuming in data preparation.

Machine Specification Files   The relevant performance paramters
of any commercial machine can be put on file for reference by the
main simulation program.   If selected, a machine is then used as
the basis for costings.

Four machine parameters are considered essential and a number of
others desirable.   The essential ones are clamp force, screw
diameter, maximum shot-volume and screw length/diameter ratio.
On the basis of these feasibilty and costing estimates can be
made.

Other relevant parameters are heating power, drive power, dry-
cycle time, melt preparation time (maximum shot), platen
dimensions and Machine Hour Rate (MHR).   If required, sensible
values are estimated during simulation in the absence of actual
data on the file.

A COMMUNICATIONS SYSTEM FOR INTEGRATED ROBOTIC ACTIVITY

S. Ashmall*

<u>INTRODUCTION</u>

Nowadays production processes in all industries are becoming more complex as they make use of developments in electronics. Robots, machine tools and sophisticated chemical analysis equipment are all employed as production tools. However, we still need simple things like on/off switches, tempera-ture sensors, and so on. The problem is to make the simple devices and the complex devices work together so that they form a complete integrated pro-duction system.

Usually the complex devices, which we can define as those devices which contain microprocessors, have some facilities for interacting with those simple devices which have been foreseen as essential to the manufacturing process. However, unless they are all obtained from one manufacturer, they will not interact easily with one another. For example, it is rare for robots from one manufacturer to interface easily with the products of another robot company. Also, it is not possible to interface to even simple devices which have not been anticipated by the robot designer.

The system which this paper describes is an attempt to provide a compre-hensive answer to all of these system integration problems. The CONNEL system is a complete kit of hardware and software which allows a system builder to integrate all the elements of his production process with minimum investment in time to learn how to use the kit. This leaves him free to concentrate on his own production machinery and the pattern of interaction which he wants to produce. Because the system builder uses CONNEL directly, he can construct exactly the system he needs. Also, when the inevitable changes to the process arise, they can be accommodated without going back to a specialist systems software house. CONNEL puts system building capability firmly in the hands of the user.

*National Engineering Laboratory, East Kilbride, Glasgow, Scotland

OVERVIEW OF THE PROBLEM

In developing CONNEL, we have been concerned to solve four main problems.
These are:

a   Making it easier to generate the complex sequences of commands which are
needed to make robots etc, perform their tasks.

b   Providing a simple way to co-ordinate such devices, so that they can act
either together or in some definite sequence.

c   Developing a way of including 'dumb' sensors, like thermocouples, so that
the actions of the complex devices can be varied in response to external conditions, such as potential error situations.

d   Aiding system construction by making it easy to test the system interactively, and then make required changes quickly and easily.

The underlying ideas of CONNEL will be described, and they will then be
illustrated by showing how they solve each of these four problems.

A REAL-LIFE EXAMPLE

To show that the illustrations are realistic, they will refer in some detail
to an actual application which we have at NEL - the automation or the lay-up
of a fibre-reinforced plastic car wheel. This application is still undergoing further development, but already involves the co-ordination of several
robots and sensors. So far CONNEL has been used for:

a   Integrating the actions of three different types of robot (Puma, Wickman
WM500 and Tosman).

b   Using a speech recognition unit to give voice control of robots.

c   Analogue sensing of the robot drive currents to provide overload
protection and touch sensing.

d   Digital sensing of switches etc, to allow such user facilities as block-
by-block operation.

e   Digital output to control robot arm power and safety circuits.

f   As a complete control system for a stepper motor robot.

These applications will be used to show how the principles of CONNEL work,
and to demonstrate that it is possible to bring almost any group of machines
under central unified control.

THE BASIC IDEA OF CONNEL

At NEL we have been involved in a variety of systems control projects over a
number of years. It has become evident that a comparatively small number of
software facilities are required over and over again. The fundamental
instructions of a computer are at too low a level for system building, and
require too much effort and programming knowledge, while high level languages
like Fortran do not have commands which are appropriate for system integration. We have found ourselves creating again and again simple programs to
do things like sending a stream of characters to a device like a robot, or

reading a character from a device, testing it and performing some sequence of actions which depend on the result of the test. To get rid of this repetitive and unproductive effort once and for all, we have put all these programs into CONNEL as simple one-word commands, and provided a simple and efficient way of combining them in any way that is needed by the application.

It turns out that about 50 primitive commands are enough to let us talk to any device. Each command is very simple. Here are some typical ones:

OUT   Output a string of characters to a selected device

CHR   Add one character to a string under construction

EQ   Test two values for equality

IFTRUE   Execute the following subroutine if last test was 'true'

LOAD   Take commands from disc rather than from operator console

START   Initiate a subroutine as a parallel activity.

To construct anything useful from these, another command is needed to link them together. This is provided by the 'DEF' command which allows any combination of the primitive commands to be linked together and given a name. The name then becomes just like one of the primitive commands and can be used to trigger the complete sequence of commands. Further, the composite command can itself be used in further definitions, and so simple one-word commands can be constructed for almost any application, however complex.

Some examples of composite commands that have been built up and used at NEL are:

MOVE   Move a robot to a specified point (this command is the same for all
           the robots we have, even although they all have different control
           languages).

SAVEL   Save all the points for the current robot to a disc file.

FINISH   Wait until all parallel activities are completed.

LAYUP   Do a complete layup of the FRP wheel.

<u>TAMING COMPLICATED DEVICES</u>

Most factories have some piece of equipment that only one person knows how to operate. It perhaps needs a lot of button pushing and typing to get it to perform, and only one person has ever had the patience and determination to master it. Thereafter there has just never been time to train anyone else to operate it. CONNEL provides a way of building the expert's expertise into a simple routine which can be given a name and called up at will by a non-expert. All that is necessary is to identify the different processes the device has to perform and define each of them in terms of the character strings or the signals sent to the machine and the responses expected back from it.

For example, suppose we had a machine that performed some kind of analysis consisting of three separate phases. The set-up for the analysis might be started by typing 'ANALYSE' on a console; then when the machine is ready, it types 'READY'; the operator responds by pressing a 'PHASE1' button; then the machine types 'READY' again and the operator presses another button marked 'PHASE2' and similarly for PHASE3. In CONNEL this could be done by typing in the instructions:

```
DEF TEST
        'ANALYSE' TOBUFF OUT       $ set up the analysis
        IN 1 SIGNAL -1 SIGNAL      $ trigger phase 1
        IN 2 SIGNAL -2 SIGNAL      $ trigger phase 2
        IN 3 SIGNAL -3 SIGNAL      $ trigger phase 3
;
```

The first line gives the whole procedure the name 'TEST'. The second line puts the word 'ANALYSE' into a buffer and sends it to the machine. The third line waits for a line to come in from the machine and then switches signal line 1 on and then off. Similarly for the fourth and fifth lines. The ';' in the sixth line just marks the end of the procedure 'TEST'. The part of each line after the '$' is free space which can be used for comments which explain what the procedure is doing. To initiate the entire sequence of typing and button pushing, the unskilled operator need only type 'TEST'.

Of course this is not an actual application, but to make this description of CONNEL more concrete and to illustrate the thinking behind the system, three more realistic examples of the use of CONNEL to interface to real-life equipment are included as an appendix. The precise details of the commands do not matter very much. What is important is to demonstrate that simple groups of CONNEL commands can generate the proper control sequences for robots and other devices used in manufacturing. The first example in the Appendix indicates the way in which a common robot language is achieved, by showing how the command to set the robot speed can be defined to set the robot speed for two very dissimilar robots. The second shows CONNEL controling an entirely different device - a speech recognition unit. The third demonstrates how a local area network can be treated as a CONNEL device and switched to different addresses.

CO-ORDINATING SEVERAL OPERATIONS

In manufacturing or process control, a common requirement is that several independent operations are started off, run at the same time, and are eventually all completed before going on with the next stage of the process. In CONNEL two ways of doing this are provided.

For situations with response times greater than a few milliseconds, a CONNEL system can run several separate tasks on a time-sharing basis. Each task is defined and given a name, just as 'TEST' was in the simple analysis example in Section 5.

It can then be initiated using the 'START' command and will run independently of any other tasks. Having got several tasks all working away, there are several ways of checking that they have all finished, so that the next part of the operation can begin. The simplest one uses the fact that the number of active tasks is available to the CONNEL user and can be tested. The command 'FINISH' is provided which simply tests this number and prevents execution from going any further until it is back at one. In more complex parallel operations, the user can set up his own flags to indicate which processes have completed and which are still running, and can allow processing to carry on in a selective way.

As an example, suppose we had three different analysis machines, like the
one in the last section. We could set up pieces of CONNEL to run each one
through its own test sequence, calling the sequences 'TEST1', 'TEST2' and
'TEST3'. When we type 'TEST1', machine number 1 will go through its test
procedure; 'TEST2' makes machine number 2 perform its special analysis; and
similarly for 'TEST3'. If each test was dependent on the previous one, we
would construct a complete test procedure as:

DEF TESTALL TEST1 TEST2 TEST3;

'TEST2' would not start until 'TEST1' was complete, and 'TEST3' would wait
for 'TEST2'. However, if 'TEST1' and 'TEST2' are actually quite independent
tests whose results feed into 'TEST3', the whole process can be speeded up by
letting them run together. This can be set up by:

DEF TESTALL START TEST1 START TEST2 FINISH TEST3;

In this way the production flowchart of a manufacturing process can be translated directly into a set of CONNEL procedures which will start and stop
activities in the correct order.

There is a second way of co-ordinating simultaneous processes. CONNEL
systems are comparatively inexpensive, of the order of a personal computer.
So a large, complex process can be broken up into sub-processes, each with
its own CONNEL controller. This simplifies the task of getting the whole
process to work, and each sub-controller plugs straight into a master CONNEL
controller, just as a complicated machine would. The master controller can
start processes on each of its sub-controllers and then wait for each of them
to run independently to completion.

WORKING WITH SENSORS

Very few manufacturing processes consist only of microprocessor-based
machines like robots. In the real world of a factory, relays, proximity
switches, and analogue voltages from temperature sensors and other sources
all play a part. The basic CONNEL controller does have sensor capability -
it has 32 parallel I/O lines as well as five RS232 ports for the intelligent
devices. For a large installation this can easily be increased. The system
of computer hardware on which CONNEL has been implemented has been chosen
because of its expansion capabilities. Additional boards can be obtained to
give analogue inputs and to provide extra parallel capability. Each of these
inputs can be sensed and used to control the complex devices. For example,
the test procedures defined in Section 6 might only be used in certain
circumstances, say when the operator has called for them by setting
particular switches. So we might only want 'TEST1' if switch 29 was set and
'TEST2' if switch 33 was on. This would be set up as:

29   RDLIN IFTRUE TEST1
33   RDLIN IFTRUE TEST2

where 'RDLIN' reads the specified input line and sets a flag 'true' or
'false' depending on whether the line is high or low. The command 'IFTRUE'
skips the following instruction if the result of the test was 'false'.

Sensors will not necessarily be close to the CONNEL controller, and it is
unwise to take signals, especially analogue signals, through an electrically
noisy factory environment. This problem is overcome by taking the signals
first to a stripped-down version of the CONNEL controller which is connected

to the main controller by a low-cost local area network to give noise
immunity. This sensor sub-system feeds back the state of digital-and-
analogue inputs on demand from the main controller and sets output signals
when ordered to do so. In this way, even remote signals can be used to sense
the state of the production system and contribute to its smooth
co-ordination.

SETTING UP A NEW PROCESS

System integration is essentially a messy business, tying together all sorts
of different equipment so that it operates in a smooth and harmonious way,
just as if it had been designed that way in the first place. To make this
possible, it is essential to be able to check out both the machines and the
instructions we give them. It is all very well to give examples of procedures written in CONNEL and say that these will drive devices in a factory.
In the real world, things are rarely as straightforward as that. First of
all, we are all human and we overlook things and make mistakes. Secondly,
the people who write the manuals for the equipment we use are also human thus
error-prone. CONNEL will work in an interactive mode which greatly aids
testing of a new procedure or even checking that a machine works as the
manual says it does.

   As an example of the ease with which a CONNEL program can be debugged, let
us look again at the 'TEST' procedure of Section 5 to drive a hypothetical
analysis machine. The procedure was:

DEF TEST

```
        'ANALYSE' TOBUFF OUT       $ set up the analysis
        IN 1 SIGNAL -1 SIGNAL      $ trigger phase 1
        IN 2 SIGNAL -2 SIGNAL      $ trigger phase 2
        IN 3 SIGNAL -3 SIGNAL      $ trigger phase 3
;
```

This might have been written by referring to the instruction manual which
came with the machine. However, such a procedure would be unlikely to work
properly first time. CONNEL allows each part of it to be executed from the
operator keyboard and the results tested for correctness. Here is a sample
session to test the operation of the procedure. On the left is what the
operator types; to the right is what CONNEL prints back.

```
OPERATOR                    CONNEL

'ANALYSE' TOBUFF OUT                        (send the set-up message)
RXTST                       6               (test length of answer)
IN                                          (read in the answer)
TYPE                        READY           (print answer on console)
1 SIGNAL                                    (check signal on 'scope)
-1 SIGNAL                                   (check signal on 'scope)
......                      .....           ......
......
```

The first line causes the initial line 'ANALYSE' to be sent to the machine,
then 'RXTST' tests to see how many characters the machine responded with.
We would expect this to be 6 - the 'READY' and a carriage return. If this is
not so, we could look at the response, if any, and check that 'ANALYSE' is
the correct instruction (and not, say, 'START'). If the response is the
expected length, it can be printed out with 'TYPE' and checked for

correctness. Then the push-button signals can be checked using an
oscilloscope or a logic probe, and so on, right through the procedure.

<u>CONCLUSION</u>

CONNEL is designed to build control systems. It puts total system integration capability into the hands of the system engineer who may not be a
computer specialist. The best way to think about CONNEL is as a kit of
simple but carefully chosen software and hardware building blocks. With some
time and intelligence, it is possible to construct almost anything out of the
basic building blocks. In addition, like any construction kit, the user is
supplied with working examples which can either be utilised as they stand or
modified to suit particular purposes. In this way precisely the control
system that is needed can be obtained without going to the expense and effort
of developing it from scratch.

The development of CONNEL has shown that certain fundamental facilities can
and must be provided in an effective control integration system, and these
facilities combine to form a list of features which can be used to evaluate
any method of control system building. Some items on the list of essential
features of any control system building kit are:

1	Universal application –	Must be able to adapt to almost any machine without requiring it to be modified.
2	Ease of use –	When adapting to a new machine or translating a new production flowchart into action must be reasonably quick and easy.
3	Low cost –	Must be economic to install and maintain. Fully automatic control systems must have high availability so complete or partial replacement is preferable to field repair.
4	Generality –	The integration strategy of the system must be general enough to accommodate machines which have not been designed yet, so that they can be used with existing systems.

The CONNEL control system satisfies all these requirements at a very
competitive price.

The field of system integration is still at an early stage, although the
need for system integration tools is becoming more urgent. There will thus
certainly be many diverse attempts throughout the market to solve the
problem. The existence of CONNEL now, with its comprehensive features may
provide a standard which must be achieved by every system offered, especially
those costing significantly more to install and run.

<u>APPENDIX</u>

SOME DEVICE INTERFACING EXAMPLES

Example 1

We use several different types of robot in our rig to produce an FRP car
wheel. To ease the problems of working with such dissimilar robots, we have
used CONNEL to create a common robot control language. A representative
section of program in our common language might be as follows:

```
PUMA                       $ select Puma robot
70 SPEED                   $ select 70 per cent of full speed
P1 MOVE                    $ move to preset point P1
WICKMAN                    $ select Wickman robot
80 SPEED                   $ select 80 per cent of full speed
PICK-PT MOVE               $ move to point PICK-PT
IF SENSE1 HIGH RECOVER     $ if faulty, do recovery.
```

   Obviously both speed commands must have the same effect. However to set up
the speed the Puma would require the character string

'SPEED 70'

followed by a carriage return character, while the Wickman needs the sequence

'M2U80'

without a carriage return. In CONNEL this is accomplishyed by creating a
separate speed definition for each robot. For the Puma it is

```
DEF SPEED 'SPEED' TOBUFF 3 BINASC TOBUFF 13 CHR OUT ;
```

and for the Wickman

```
DEF SPEED 'M2U' TOBUFF 2 BINASC TOBUFF OUT ;
```

   'TOBUFF' transfers a string of characters from a temporary internal buffer
to the device output buffer, 'BINASC' converts the speed value provided into
a character string, and 'OUT' outputs the constructed device output buffer to
the currently selected device. Both commands have the same name but are
correctly obeyed for each robot because they are stored in separate
vocabularies which are hidden or revealed by the use of the robot name.

Example 2

The speech recognition unit we have has an initialisation sequence consisting
of the three lines

```
          'T10000' (set status)
          '3'      (reset the unit)
          'HC'     (start loading vocabulary)
```

   Each line is preceded by the character CONTROL/B (which has the numerical
value 2) and followed by line feed (= 10). A single command to do this is
'INIT' defined as:

```
DEF INIT 2 CHR 'T10000'  TOBUFF 10 CHR OUT IN DROP
         2 CHR '3'       TOBUFF 10 CHR OUT IN DROP
         2 CHR 'HC'      TOBUFF 10 CHR OUT IN DROP
;
```

The 'IN DROP' part is just to ensure that we wait until the unit responds before we send it the next line.

Example 3

As a final example, we will look at switching a local area network node to talk to a new station. The particular type of network node we use accepts CONTROL/A (= 1) as a 'reconfigure' command. It replies with a line of 51 characters showing the current configuration and askng if this is acceptable. We reply with the character 'N' for 'no'. The node then accepts a new destination number of two numerical characters followed by carriage return (= 13) and gives us back 48 characters confirming the change. To break the problem up a little, all of this is programmed as three sub-commands WFCONFIG, WFCONFRM and NODE, which are combined into the actual command to switch the node as follows:

```
DEF WFCONFIG RXTST 51 LT ;
DEF WFCONFRM RXTST 48 LT ;
DEF NODE      IOINIT *$ NODENO 255 - ! 1 CHR OUT
              WFCONFIG UNTIL IOINIT
              'N' TOBUFF 13 CHR OUT
              2 BINASC TOBUFFG 13 CHR OUT
              WF CONFRM UNTIL IOINIT
              99
;
DEF SETNODE DUP DUP NODENO EQ 1 SWAP -
            IFTRUE NODE DROP
;
```

Details of the commands are less important than the demonstration that it is possible to interface to a specialised device with comparatively little work and without any hardware modifications at all.

Part III

Competitive Manufacturing Systems in Continuous and Discontinuous Processes

AUTOMATION OF INJECTION MOULDING—AT ANY PRICE?

H. Eckardt*

Paper not available at
time of publication.

* Battenfeld Maschienfabriken GmbH, FR Germany

PANEL PRESENTATION: SOME EXPERIENCES IN ADVANCED MANUFACTURING TECHNOLOGY

R. Austin, T. Zavos, and J. Czerski

<u>Robotized product handling</u> (R Austin*)

Automated handling of moulded products directly from an injection
moulding machine produces:

1   increases in productivity

2   improvements in product quality

3   a catalytic effect upon professionalism

It will be illustrated by reference to actual case studies that
the introduction of robotized product handling has a direct
effect upon the quantity of products produced, over even a fairly
short period of operation although the 'button to button' cycle
time may be longer than that best achieved by human machine
attendance - continuous production as opposed to continual
production.

With reference to actual case studies again, the influence upon
improved product quality by resultant continuous production is
demonstrated to be quite dramatic. The analysis is made that a
cyclic thermal process requires to operate in a time constant
environment.

Finally the unexpected catalytic effect upon professionalism which
most often subsequently descends around the robot equipped machine.
Often initially perceived by the proud new robot owner as the
panacea, it is rapidly discovered that the lack of 'human flexibility'
of the robot prompts attention to areas of shortcoming elsewhere
in the production process. The ever present desire to 'do it right -
tomorrow' suddenly becomes an immediate demand to 'do it right -
today'.

*Pressflow Ltd

A case study of introducing advanced manufacturing technology

(T Zavos*)

The work discussed in this paper was initiated at Elco Plastics
Limited in High Wycombe, a division of Thorn EMI Ferguson. Indeed
part of the exercise was the relocation of the factory, still
within High Wycombe, providing a more suitable environment for a
high technology plastic moulding company.

The picture at the start of the project was one I am sure familiar
to all of you, a traditional trade moulding environment. The
unfortunate reality was, though, that market forces were leading
the company to an avenue to which its current methods and practices
could not respond.

It was necessary to be much more flexible and to drastically reduce
costs. At the same time, the market required products of low
material cost content but of extremely high quality finish –
precisely moulded components, satisfying complex engineering
requirements, finished in two or more colours and printed with
complex logos or cosmetic designs.

In short, we felt that we could translate the new market
requirements in the following terms:

(1)   Sophisticated process control

(2)   Stable cycle times

(3)   Increased production output through consistent operating
      conditions

(4)   Accurate forecasting of production outputs

(5)   Reduction in scrap

(6)   Reduction in labour costs

(7)   Improvement in materials handling methods

There is nothing really earth-shattering in the above objectives,
but our next task was how to set about achieving them.

As a company, we felt that robotics would be a good start, to help
us slowly to build the kind of environment conducive to the quality
levels needed in the market place.

The two most problematical areas were moulding and spraying and these
were the first two areas to be attacked with robotics in mind. An
initial attempt involving two pick and place robots and two spray

  * Thorn EMI Ferguson

robots was set up and closely monitored. The evaluation was done
in full production conditions and with no relaxations of day to day
production requirements. This presented us with substantial problems
initially but it was the best way to learn what robotics could offer
us. More to the point, it was also the best way to find out what we
wanted the robots to do for us.

The second stage of the project for us was to ensure that our
experiences were put to good use. We had a requirement for a total
robot population of more than 25 pick and place robots and 4 spray
robots. We spent a lot of time raising a very tight specification
reflecting our experiences, good and bad. With the specification in
hand, we then went through a normal selection process involving most
European makers and we eventually made our purchase.

In the case of the pick and place robots we had to cover machines
varying from 250T to 1500T capacity. We achieved this using robots
from one supplier, to identical technical specification right down
to wire numbers used on interfacing. The moulding machine layout
was developed with robots in mind and this has enabled us to easily
run two machines using one operator. Second operation work is also
possible now within cycle time. The ability to do such things as
product weight checks within cycle time has given us added control
of the process.

It would be wrong for one to assume from the above that the kind of
improvements we were looking for could be achieved through the use
of pick and place robots alone. Some of the other measures taken
were:

(1)   Automatic powder feed system

(2)   Closed loop process control of moulding machines

(3)   Correct machine layout

(4)   Automatic production output monitoring system

When considering the use of spray robots we had to deal with a great
variety of product sizes and finishes from simple, single colour,
television front-frames to two colour frames with difficult break
lines and large 22" one-piece television cabinets. This robotic
application further demonstrated to us the importance of attention
to peripheral detail. The robot in this case was only a small
problem compared with that of designing a 'clean room' type
environment, for high quality spraying. Conveyors, paint supply
systems and health and safety requirements contributed equally to
the sleepless nights!

It is indeed currently the case within industry that often robotic
solutions are adopted for the wrong reasons. It is, in fact, even
more often true that blame for failure is conveniently levelled at
the robot when the true blame lies elsewhere. Some of the real
problem areas are:

(1)   Incorrect specification

(2)   'The Managing Director said we must have robots' syndrome

(3)   Product not suitable or not correctly designed for automation

(4)   Badly designed grippers and feeders

(5)   Lack of correct environment for robots.

Capital expenditure is yet another thorny problem which the potential robot user has to tackle. In our case we found it not extremely difficult to justify our expenditure through conventional means of scrap reduction, labour cost reduction and improvements in efficiency. One has to exercise great care when cycle time reductions are mentioned. They are usually due to the higher disciplines demanded when using robots rather than directly due to the use of robots. In other words, these improvements could possibly be achieved without the robot and through better setting, better supervision, better maintenance, etc.

I do however believe that with more and more expensive robots being used in industry, conventional means of justification can no longer assist. Flexibility, for example - something extremely important to all of us these days - is a concept which is not particularly financially attractive to our colleagues in the Accounts Department. Finally, and perhaps controversially, justification may be the fact that some of our potential customers are demanding that their product is handled by 'robot user' companies.

<u>Materials and advanced manufacturing technology</u> (J Czerski*)

As is well appreciated by most convertors, the consistency of
performance of a polymer is not only dependent on its chemical
form and molecular weight, but also on its internal structure
as generated in the moulded form. This in turn is dependent on
the consistency/reproducibility of the processing conditions used.

Experience has shown that in industrial failure analysis and
process optimization the microscope is a valuable and widely
applicable instrument. Using this technique the author will
attempt to illustrate how modern methods of machine automation
can lead to improvements in grade choice and consistency of
the product produced.

*   BASF UK Ltd

ADVANCED MANUFACTURING TECHNOLOGY IN THERMOSET MOULDING

P.J. Clarke*

In the area of Thermoset injection moulding machine investment moulders can be faced with a deluge of information, particularly from machinery manufacturers. He is then left with the difficult decision to balance investment with production costs and future prospects. By selecting a machinery manufacturer who can offer continuous development of processing techniques together with CNC control systems, telediagnostic trouble shooting, production managment systems and material handling systems, the equation falls heavily on the side of investment.

INTRODUCTION

In recent years the application of Thermosetting materials has once again made its mark on the Plastics industry. This is particulary true for the case of Glass reinforced Polyesters where their useage has increased several fold for many exacting applications. Not least of these being the Automotive industry.

From the first developments for DMC, BMC and SMC, Bucher-Guyer have been in the forefront of processing developments with specially designed "Valve Ring Assemblies" for the elimination of material reverse flow and "Stuffing Units" for material handling.

The development of these "Valve Ring Assemblies" for DMC has led to further design and useage for more conventional Thermosetting materials. This equipment enables even closer control of "Holding Pressures" and "Shot Weights" than hitherto possible.

THERMOSET INJECTION MOULDING MATERIALS

Generally all Thermosetting materials can be moulded by the injection moulding process.

These include :–   Phenolic
Melamine
Urea
Epoxy
Granulated Polyester
Unsaturated Polyester
(DMC,BMC,SMC,TMC)

* General/Sales Manager – Bucher-Guyer (UK) Limited

Raw materials suppliers originally developed these materials for the compression moulding process, but with the advent of injection moulding they were quick to respond to the industries requirement.

The main objectives to achieve these requirements were:-

- longer standing times, even with high
  cylinder temperatures.
- good flow behaviour
- fast reaction to tool temperatures
- improvement in wear characteristics.

Today these so-called injection moulding properties are available for almost any thermosetting material.

ENTER THE POLYESTER ERA

The major growth area for thermosetting plastics is undoubtably in Unsaturated Glass reinforced Polyester. All of the following come under this general heading. DMC (Dough moulding compound), BMC (Bulk moulding compound), SMC (Sheet moulding compound) and TMC (Thick moulding compound).

Polyester material has probably the shortest history of any thermosetting materials, but nevertheless has attracted the most interest, especially from the electro-household, automotive and electrical industries.

The great advantages of polyester materials, especially over thermoplastics are those attributed to standard thermosetting materials. ie:-

- high temperature resistance
- high electrical resistance
- rigidity

In addition, glass reinforced polyesters have greater strength and can produce high finish cosmetic parts in an infineitly wide variety of colours.

Machine requirements

In the early days of injection moulding of polyesters two types of injection units were evaluated. These being the Reciprocating Screw and Plunger or Piston type. It soon became clear that despite the very slight reduction in component strength attributed to screw plasticizing, the advantages of this system precluded all others. It is important to note that this slight reduction is only evident when using materials with particularly long glass fibres.

The advantages of the screw preplasticizing unit are:-

- better homogenity
- superior surface finish
- closer control of shot weight
- reduced gassing

However in order to fully achieve these advantages it was necessary to employ some method of preventing the polyester material from flowing back up the flights of the screw. This being necessary because of the low viscosity of the material. The answer to this problem was the Reverse Flow Lock, perhaps better known as a check ring or valve ring assembly.

Screws - In order to achieve minimal degredation of the glass fibres through the screw, a different design from that used for conventional thermoset materials was required. This culminated in a design of screw with greater depth of flights and less compression, coupled with the use of special steels and complicated hardening procedures. The Polyester screw is now a standard feature in the range of options for processing thermosets.

Valve ring assembly - Conventional valve ring assemblies such as those used in processing thermoplastics consist of a tip (valve body), ring and pressure pad. The tip (valve body) is in the form of an arrow head, generally with 3 flutes. Until recently this type of design was used in the processing of polyesters and still is by several manufacturers. However, the latest design employs a plain tip (valve body) and sculptured valve ring. As unsaturated polyester is sensitive prior to curing, this new design eliviates the problem of a material seizure around the arrow head.

Stuffing units - Because the very nature of unsaturated polyester in its raw state is that of a "putty" consistancy, the requirement for special handling and feeding is essential. The stuffing unit is used to force the raw materials through the barrel feed throat on to the screw.

Storage and automation - In order to produce a complete "fully automatic cell" it became necessary to develop a method of handling the raw material within closed containers, this is to prevent the styrol in the material from escaping.

This is achieved by using a loading device known as a "Poly Loader". This consists of 12 drums, each containing batches of DMC or SMC of 20-50 kg each. The drums are mounted in a circular formation so that each drum is located in turn, with the stuffing unit. Thus allowing the material to be fed automatically from the Poly Loader into the machine.

Processing - Accuracy in the processing functions for Polyester are of special importance because of its critical nature, specifically in respect of temperature. As the material leaves the stuffing unit and enters the barrel it has a temperature of approximately 20 deg. C. As the material is transported down the screw, the temperature is gradually increased until it enters the nozzle at between 40 - 50 deg.C. By using an extended temperature controlled nozzle, it is possible to maintain the material at a constant temperature until it is deep within the mould, thereby reducing the length of sprue and with thin walled parts, reducing the overall cycle time.

In order to achieve optimum surface finish it is necessary to inject the material at high speeds. This however, has the effect of reducing the overall part strength.

Typical injection moulded DMC components include:-

- telephone distribution boxes
- copying machine housings
- sandwich toaster parts
- circuit breakers
- fuse boxes
- head lamp reflectors

CLASSICAL THERMOSETS

<u>Screws</u>- In the area of classical thermosets as outlined on page 3 there are several alternative screw designs :-
ie. - for epoxy - without compression.
    - glass filled Phenolics - increased flight width and special screws for the manufacture of commutators and electro motors where extremely high injection pressures are required for very stiff flowing materials.

<u>Valve ring assembly</u> - One of the most recent developments for the processing of classical thermosetting materials is the "Reverse Flow Lock" or "Valve Ring Assembly". Where standard thermoset materials, particularly when those with a low viscosity are used, similar problems occur to those experienced with unsaturated polyester, ie.- reverse flow of the material during the injection stage. Encouraged by the excellent results achieved with DMC screw developments, a new screw with valve ring assembly has been developed. In comparison to the DMC screw, the flight depth is less and the design, steel and hardening process of the valve ring are different. This unique facility produces minimal reverse flow of the material, resulting in less screw and barrel wear, a higher utilization of the available shot weight, less shot to shot variations, better homogenization of the materials, a better defined injection and holding pressure profile and therefore a better finished component. The following table shows the relative material utilization, comparing components produced with and without a valve ring assembly.

TABLE 1

Material	Component	Shot Wt.	Plast. Stroke		Check Ring	
					with	without
PF Type 31	Contact Housing	70.5	45	(60)	90	67
MF Type 152.7	Pan Handle	233	138	(165)	90	75
MP Type 182	M/C Model	286	152	(185)	94	77
GR.UP.	Cover	267	80	(95)	94	79
Units	-	g	mm	(mm)	%	%

() without check ring assembly.

The % values refer to the ratio between the plasticized material and the effective injected material.

- The benefit of the check ring over the standard screw tip is around 15%

When processing with the valve ring assembly the homogenity of the material is improved causing any air in the material to be pushed backwards. Thus the temperature consistancy is improved resulting in shorter cycle times. Generally it can be said, that the thicker the component, the more effective this process becomes, which consequently produces even more savings on the overall cycle times.

SPECIAL PROGRAMMES

In processing thermosetting materials, special programmes are often required. Taking these in turn, they include:-

Breathing - generally associated with thermoset materials this programme is used to remove air or gasses within the the mould. Once the mould is closed and has been filled with material it is allowed to open again by a small accuratly controlled amount and allow the escape of any gasses which may have formed during the initial curing process

Injection-compression - this programme is used for components which would typically be moulded by the compression process. The main advantage being, that the finished component has a higher strength than would be produced by straight injection. The sequence of operation for injection-compression, is that the mould is partially closed to within approximately 5-10 mm of the fully closed position. At this point a predetermined quantity of material is injected into the cavity. When all the material has been injected the mould is fully closed, forcing the material, by compression into the form of the finished component. This method of moulding provides advantages attributed to both compression and injection moulding.

Vacuum extraction - this programme is particularly useful when moulding DMC. By manufacturing the injection mould with a seal around the cavities, the mould can be evacuated from air prior to injection. This process eliminates reject mouldings caused by air traps and gassing.

CONTROL SYSTEMS

In general the industry has gone through 3 changes in the type of control systems available. These being:-

- Relay
- Digital
- C.N.C.

Relay control - has almost ceased in manufacturing today. It is very slow in operation and extremely difficult to expand.

Digital control - came in either a dedicated logic system or more popular, the programmable logic system. This type of control allows all the machine parameters to be set from a machine control cabinet. These include distances, times, pressures, speeds and temperatures. The sequence programme for the programmable logic system can be changed or added to by reprogramming the memory, usually in the form of "EPROMS". Thereby making the system flexible and expandable.

C.N.C. - is without doubt the most flexible type of control system available to date.

By presenting the operator with information in page format, only one set of values are available for adjustment at any given time. This method of setting the machine parameters, whether for example they are barrel temperatures or injection speeds is done in a logical manner by providing the operator with simple and precise alternatives.

With the use of a data recording device or memory, all the machine parameters appertaining to a particular moulding tool can be recorded. This greatly reduces the time required for resetting a mould for subsequent production runs and generally means a mould can be back in production within minutes of achieving the correct moulding temperatures.

Trouble shooting – With CNC control trouble shooting becomes the domain of the setter rather than that of an electronics expert, as generally 90% of all failures fall into a short list of troubles. For example these would include :-

> - safety gate not closed
> - ejectors not returned
> - out of material
> - etc. etc.

All information appertaining to these failures is presented to the operator on the vidio screen for rectification.

Telemodem – As an extension of the standard trouble shooting facility the use of a Telemodem enables the machine user to connect directly with the machine manufacturer by means of the normal telephone system. Thereby the electronic signals are transmitted acoustically, enabling the manufacturer to trouble shoot or reprogramme without actually being in attendance at the machine.

Production management systems – Moulding companies who have installed Production Management Systems have realized the benefits of planning, statistical analysis and production control, thus achieving a far higher rate of machine utilization than would otherwise have been achieved. Indeed, machine efficiencies in excess of 95% are not uncommon by the use of high quality equipment and close production monitoring.

CONCLUSION

By the use of modern technology employed in machine design, coupled with advanced electronic control and Production Management Systems todays moulder can quickly and efficiently recoup his captital investment, in addition to realizing many more moulding prospects than would otherwise be possible.

We must therefore conclude that companies who invest in modern plant and technology will undoubtably reap the best and most profitable fruits.

ADVANCED MANUFACTURING TECHNOLOGY IN BLOW MOULDING: THE NEXT MAJOR GROWTH AREA?

Robin Enderby*

Tooling for blow moulding is usually cheaper than that for
injection moulding, and the flexibility of the process
together with the increasing range of materials with
which it can be used, make blow moulding ripe for
significant growth.

The blow moulding process is frequently thought to be primarily suitable for
producing plastic bottles in polyethylene. In fact, UK blow moulders are
producing a wide variety of mouldings out of materials such as acetal,
polypropylene, PVC, polycarbonate, PET, EVA, thermoplastic rubbers, and
polyethylene from low density to ultra high molecular weights.

In addition to this wide range of materials, multi layer mouldings and
surface coatings are also used to increase resistance to solvents, or
reduce taste tainting caused by oxygen permeation.

Blow moulding machines are producing mouldings of 10cc to 10,000 litres,
4 metres long (wind surfer boards), and containers with burst strengths over
1000 psi.

Apart from the materials now suitable for blow moulding, the machinery
itself has been developed to run at cycle times more than double that of
10 years ago. Better cooling, faster proportional hydraulics and more
advanced controls have all contributed to this improvement.

Most modern machines can now produce mouldings which are fully deflashed,
standing upright, leak tested, and physically checked for moulding defects.

A large growth area for blow moulded parts is the automotive industry. Car
fuel tanks are now being moulded in high molecular polyethylene. Rubber
gaiters are now being moulded in 'thermoplastic rubbers' with improved
performance.

The advent of in-mould labelling could have a massive effect on product
appearance and on the printing industry currently using silk screen or
offset printing methods.

* Blow Moulding Controls Ltd

Although the blow moulding process cannot usually compete with injection
moulding cycle time, mould tooling is usually significantly cheaper, and
the flexibility of the process coupled with the increasing range of materials
will ensure that a larger percentage of components will be blow moulded in
the future.

Traditionally, the blow moulding process has been less automated than
injection moulding.   The reason probably lies in the very wide range of
shapes and sizes of mouldings which can be produced by the blow moulding
process.   Production runs are frequently short, which will inevitably
reduce the viability of full automation.

Because of the limited market demand for specialist mouldings it is
inevitable that there will be companies which will find full automation
impractical, and not cost effective.   Many of these companies could reduce
the labour content of their products by examining each aspect of the process
and eliminating or reducing the labour content in any area which can easily
be automated.

When considering automation it is essential to first ensure that the basic
moulding process be as stable as possible.   Automation equipment usually
expects to be presented with stable, logical parameters.   For example, a
deflashing machine can jam solid when presented with a hopelessly deformed
moulding.   Another area well known to cause trouble to automated systems
is the quality of material being supplied to the machine.   It is inevitable
that a percentage of this will be reground.   It has been found that the
biggest cause of inconsistency is caused by variable amounts of regrind, or
poor mixing of regrind with virgin.   Some major moulders have relegated
regrind to bulk, black coloured, low specification mouldings.

Once a stable process is achieved, consideration can be given to the
possible methods of automation currently available.   Diagram A shows the
typical stages in the blowmoulding process.

Examining each stage we have:-

Incoming Materials

These can be purchased in 25Kg, 1 ton or 20 ton bulk loads.   The optimum
quantity can depend not only on usage but also on how frequently a material
change is made on any machine.   Small companies should not rule out the
possibility of bulk loads if they only use one type of material.

Localised 'by the press' regrinding and mixing is now the most popular and
practical for most moulding companies of small to medium size.   This method
avoids contamination from other colours or materials, and assists in
maintaining a constant regrind percentage.   The labour required to carry
away scrap to a central regrind machine, and return the material, cleanly,
to the correct machine is eliminated.

Masterbatch colourant is expensive, and the cost of manual measuring and
mixing is also expensive.   Automatic metering systems can be set to
minimise usage of masterbatch with very low manual involvement.

Metal Detection on the Feed Hopper

Any feature which prevents breakdowns automatically assists automation.
Systems are available which reject metal particles automatically without
stopping the moulding machine.

Deflashing

Automatic integral cropping, deflashing and punching is now offered by
virtually all blow moulding machine manufacturers.   Systems are also
available for retrofitting to existing machines.   Because of the wide
variety of moulding shapes, changing over deflashing equipment can presently
be a time consuming job.   Some machine manufacturers have adopted a modular
approach to this tooling which is at least a step in the right direction.

It may still be cost effective to use manual labour for loading deflashing
machines, as depending on the size and type of moulding, one operator can
look after the output of two or possibly three machines.   This is made
possible by the deflashing machine having located the moulding in a 'datum'
position, allows for a 'pick and place' robot, or palletising machine to
stack or pack, which saves the operator this task.

This 'hybrid' method of partly automating simple tasks,   and leaving humans
to the jobs for which machines are not presently cost effective, is
probably the best approach for 'trade moulders'.

Automatic Testing

There have been automatic leak testers and dimensional testers in use for
many years now.   There has been an increasing interest in automatic leak
testers with the filling companies worry about legal liability from damage
caused by leaks, also the mess and downtime caused on the filling lines
by leaking containers, is now unacceptable.   Most major UK blow moulders now
have leak testing equipment.   The viability of this equipment and dimensional
testing again is dependant on the moulding being held oriented, deflashed in
a 'datum' position.   Operator hand feeding of this equipment is not an
efficient use of human labour.

Computerised video analysis equipment which can check dimensions, colour,
contamination particles, and completeness of moulding is still considered
far too expensive, and has not yet reached a 'state of the art' level which
moulders are looking for.

Automatic Mould Changing

It is standard practice in blow moulding never to make two moulds the same
size.   The cavities are surrounded with just enough metal to bolt the
mould to the platen.   The injection moulding industry has made moulds
from standard sized plates for years.   With a little careful planning,
families of moulds can be made which lend themselves to automatic
alignment methods, and manifolded cooling waterways.   Because of a mould's
weight, it is impractical to consider moving moulds with conventional
industrial robots.   The method used in injection moulding is a railway
track system linking the back of each machine with the mould stores.   Very
impressive change times are quoted of just a few minutes or less, and for
trade moulders on short run work, such systems could quickly pay for
themselves.

<u>Finished Product Handling</u>

This is one area ripe for improvement in almost all blow moulding companies. The blow moulding process produces <u>bulk</u>. Because most mouldings are a hollow plastic bubble, the moulding process is really a conversion from a compact form of plastic (granules) to a space filling problem.

It is essential to:-

a)   Minimise the number of handling operations of the finished product.

b)   Automate any necessary handling as far as possible.

The first practical solution is to avoid using the immediate production area for storage. The congestion causes inefficiency, frayed tempers, and actually some physical damage to machinery.

Some companies convey mouldings completely away from the machine area which then allows subsequent operations to take a 'flow line' path layout. This approach can require more space than the 'look how many machines we've squeezed in my garage' system!

The moulding handling problem is aggravated by the difference in speed of finishing operations, to the speed of the moulding process. For example, offset or silk screen printing machines inevitably run much faster than blow moulding machines. Because of this, mouldings are packed into boxes, temporarily stored, unpacked and unscrambled, printed (or other finishing operations) re-boxed and moved again to a storage area ready for shipment. The costs of all this material handling can be higher than the material and moulding costs put together. Worse still, it can be particularly difficult to accurately establish the costs of all this manual product handling.

<u>Improving Efficiency by Automating Testing and Finishing Processes and Integrating them into the Moulding Cycle</u>

The ideal production is shown in figure 2.

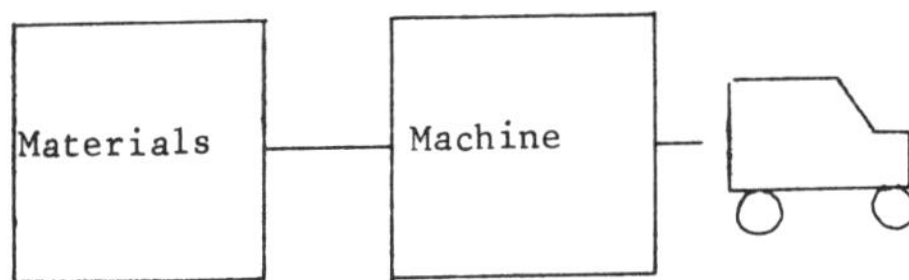

One way of approaching this utopia is to try and integrate finishing and testing processes into the moulding machine.

Currently available processes include:-

Integral leak testing

Sensing of blocked bores and bore sizes measurement

Checking for unwanted flash or excessive flash

In mould labelling (IML)

Of these, 'in mould labelling' has the potential to revolutionise both the work flow problems associated with conventional printing, and the way which containers, or other mouldings are currently decorated.

Probably the biggest stumbling block will be the packaging designers who decide what consumers are most likely to buy.   Although sometimes innovative, they are cautious in packaging many products in containers 'accepted' for its use, e.g. washing up liquid bottles are all very similar, and faced with these style limitations, a gold foil, non-slip grip, holographic image label is unlikely to be adopted for this application.   Blow moulding companies could perhaps, sway packaging designers to choose an IML decorated moulding, by providing a price advantage.   Why offer a moulding produced by a fully automated process at the same price as one produced by a labour intensive process?   The customer will have no incentive to choose products made with the benefit of automation.

One temptation could be to put a premium on products produced by non-automated processes.   This would have the effect of allowing small, low-overhead, non-automated companies a better chance of competing, particularly on short run work.

<u>'Short Run' Work and the 'Trade Moulder'</u>

It is often wrongly assumed by small companies, that advanced automation is only relevant to large companies with long runs of the same product.   This is not true, but the level and type of automation may be different.   Consider again 'in mould labelling'.   It is feasible to put a different label onto every moulding i.e. a production run of one!   Small companies can also benefit from automated testing, because of their frequently limited Q.A. and laboratory facilities, it relieves them of trying to 'rig up' something which inevitably is disruptive.

<u>The Automotive Market</u>

The biggest challenge facing the blow moulding industry is undoubtedly, the massive increase in blow moulded parts used by the automotive industry.   The range of shapes, sizes, materials, and particularly finishing operations are much greater than the packaging industry.

Because of the high quantities, duration of run, and scheduled production, it is already obvious that blow moulding machines are being specifically designed to produce individual mouldings.   The all plastic car petrol tank is an example where the moulding machine, finishing machining operations and sulphonation are fully automated.   Radiator expansion bottles are moulded in polypropylene on machines equipped for high shot weight, small size thick walled mouldings.

Carburetter floats are moulded in acetal by special techniques to prevent
the float sinking.   Steering gaiters and rubber 'boots' are now being
moulded in 'thermoplastic rubbers' which give longer service life than
conventional compression mouldings.   Spoilers, air dams, side skirts and
filler pipes are all now being blow moulded with the desired reduction in
vehicle weight and price, and in most cases, far superior performance over
the materials they replaced.

The range of materials and processes will guarantee that the blow moulding
method of producing hollow, complex shapes will form an increasingly
significant part of both private and commercial vehicles.

<u>In House or 'In Plant' Moulders</u>

Companies who mould for their own use have probably the greatest to gain
out of total automation.   It is entirely feasible that polyethylene, water,
and few chemicals enter a building, and cartons of washing up liquid leave
the building.

Cost savings can be made in time, manpower, and production control can be
simplified, if all the elements of a product are made under one roof.   It
may also be possible to save certain items of equipment, for example,
there may be no need for a bottle unscrambler on a filling line if the
bottles arrive already standing up from the blow moulding machine (bottles
IML, or course).

A car manufacturer could also benefit if mouldings came straight from the
machine directly to the assembly line, instead of a costly stores, goods
inwards, Q.A. etc. system.

It should be pointed out that not all in house blow moulding operations have
been totally straightforward, as blow moulding is known to be more 'black
art' than the science of say, injection moulding.   The obvious, and
ultimate 'in house' step is to fill containers in the blow moulding machine.
This is already being done with milk and sterile medical fluids.   Many
companies moulding 'in house' still buy significant quantities of bottles
from outside.

<u>Centralised Monitoring of the Blow Moulding Operations</u>

Better automation, faster cycle times, and less manual involvement, creates
a need for an 'automated' production monitoring system.   Standard systems
on the market are based on operators setting a 'condition status' switch on
the moulding machine, the setting of all these 'status' switches is
monitored by a central processor, which displays all relevant production
information.   Due to the rapid developments in micro processors, some
companies may now find that their production monitoring system is already
out of date.

These systems collect cycle time information, reasons for stoppages, and
stopped time, and the details of the job running.

A smart monitoring system should provide each manager with the information translated into the form most suited to his needs. For example, the financial manager might like to see utilisation, and total value of factory output. The Purchasing Department may only need to know the weight of polymer being produced and predicted usage. The Maintenance Department need to be alerted to breakdowns, but might also find it useful to see which machines are less reliable than others. And of course, the production manager should use the system for scheduling - but a smart system could attempt to match resources to requirements.

This is not as futuristic as it sounds - various industries, manufacturing and transport, have used computerised scheduling for years.

Systems which have only one computer with 'slave' monitors will probably not be capable of flexibility needed by a highly automated factory.

<u>Training</u>

Advanced automation equipment often utilises the highest levels of electronic technology. It is quite possible that in the totally automated factory, that the plant will not be maintained by 'moulders', but by electronic technicians, or even software engineers.

Training of personnel will be more important than at present, because high tech faults are not fixed by a few small adjustments and bending brackets. It should not be assumed that high tech will be ultra reliable. The pneumatic valves, cylinders, and other mechanical devices often have manufacturer rated lives of 1 to 50 million operations. This may only represent one month's output. Also, devices like pick and place robots may make 10 or more positional movements for each moulding, and this could be particularly unreliable in high production environments.

Although moulding machines now have limited self diagnostic features, the smartness of such systems is currently limited to a first order message e.g. "valve energised, but piston didn't move". It could be that the plastics training establishments will need to tailor courses specifically for electronics engineers.

<u>The Best Way to Automate</u>

Converting existing machines to full automation is not an easy task. Many blow moulding machines were designed when it was totally acceptable for a high level of manual involvement. Consequently, they do not always lend themselves easily to automation.

The ideal situation is to plan for a high degree of automation during the original factory planning stage. This inevitably, will limit total automation to the largest companies, as the inevitable capital outlay will be beyond the reach of the small company which usually starts with just one machine.

Some moulding machine manufacturers offer 'turnkey' installations, i.e. complete design layout and equipment supply. However, some of these manufacturers only have limited experience of certain types of installation such as dairies, or large plastic drums. Ancillary manufacturers, or their agents can be a valuable source of advice through their experience of applying automation to existing production lines

<u>The Future of Blow Moulding with Advanced Automation Techniques</u>

Blow moulding machine manufacturers have finally grasped the need for
machines which do more than simply blow a plastic bubble within a mould
cavity, although it would appear that blow moulding machines still lag
far behind the technology already available on injection moulding machines.

Hopefully, the next few years will show automatic mould changing and
mould 'auto set' facilities.   Automatic start up from a single button
would also be welcome - also with automatic shutdown, which would include
automatic purging of unstable materials.

Stored programme settings are already available, but this computer option
is more expensive than conventional logic, and surely this should change?
Autosetting temperature controls are now becoming available with process
or control, although few machine manufacturers are fitting them.   Integral
interface for ancillary automatic devices with extra hydraulic tappings
for local pick and place robots.   And finally the ultimate.   Control from
a remote location e.g. Production Control office to:-

    load required mould

    set correct material and colour

    auto-set machine parameters

    integral testing

    automatic warehousing

These are not pipe dreams, because this level of automation is already
available and is being used in other industries, and if you don't do it,
your competitor might!

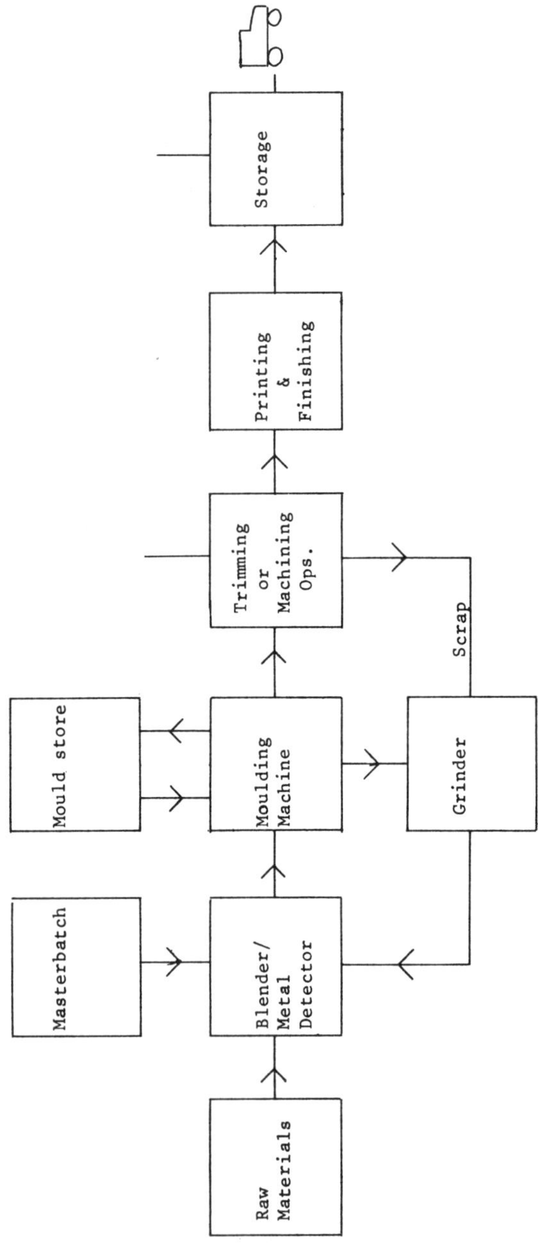

Storage
Printing & Finishing
Trimming or Machining Ops.
Scrap
Mould store
Moulding Machine
Grinder
Masterbatch
Blender/ Metal Detector
Raw Materials

Acknowledgements to companies assisting in the preparation of
this paper:-

Plysu PLC, Woburn Sands, Beds.

Bettix Limited, New Malden, Surrey.

R.B. Blowmoulders Limited, Canning Town, E16

Avon Industrial Polymers Limited, Bradford on Avon, Wiltshire.

ADVANCED MANUFACTURING TECHNOLOGY FOR
FILAMENT WINDING

M.J. Owen*, D.G. Elliman*, V. Middleton*, K. Young* and N. Weatherby*

The filament winding process is described, and
its potential for the manufacture of structures
of high specific strength explained. It is
shown that the use of computer aids for design
and manufacture is crucial and that purpose
built control systems are needed for high speed
filament winding. Several problems encountered
in winding complex geometries are identified,
and possible solutions proposed.

INTRODUCTION

Filament winding is a long established process for the manufacture
of composite materials. In the past the technique has been used
for the manufacture of simple mainly cylindrical geometries such as
pipes, gun barrels, and rocket launchers. The filament winding
involves laying down a bundle of fibres on a mandrel surface in a
precise geometric pattern. The fibres are termed a roving, and the
process has traditionally been carried out on a helical winding
machine. This tool has much in common with a screwcutting lathe,
and is shown in figure 1. The angle alpha is described as the
winding angle.

The roving may be pre-impregnated with resin, passed through a
wet out system, or winding may be carried out dry and the resin
injected in a subsequent process. The mandrel may be left in place
as a stiffener, or gas/fluid seal in some applications but must be
removed in others.

Filament winding is currently used to manufacture components with
simple geometries where high strength to weight ratios are
required, or good chemical resistance is needed. The strength and
stiffness of a component depends critically on the orientation of
fibres in the matrix, which can be carefully controlled with this
process.

In recent years the advent of fully programmable, multi-axis CNC
(Computer Numerical Control) machine tools has extended the
potential application of filament winding to complex geometries.

* Department of Mechanical Engineering, University of Nottingham

The generation of suitable fibre paths on the surface is a challenging problem for non-axisymmetric shapes, making computer aided design an essential prerequisite for the exploitation of the process.

THE NOTTINGHAM FILAMENT WINDING MACHINE

The filament winding machine used at Nottingham is illustrated in figure 2. It is a Pultrex MODWIND 1S-5NC with a single spindle and 5 independent numerically controlled axes of movement. These comprise mandrel rotation, three mutually perpendicular carriage motions and a rotary movement of the payout eye. The latter feature is used to avoid the introduction of twist into the roving, and is essential if tape is to be wound. The machine can accept mandrels of up to 3 metres long by 1 metre in diameter. The maximum speed of a linear axis is 40 m/min. and the maximum speed of a rotary axis is 100 rpm. The machine is controlled by a Bosch micro 8 CNC machine tool controller. This uses standard machine tool NC commands to position the axes, and also has a teach-in facility. The majority of the NC commands are redundant for filament winding, as they assume a metal cutting tool in contact with the surface to be machined. The pay out eye in fact describes a locus relative to the surface of the mandrel along the tangent at any point to the fibre path being wound.

GAINING THE MAXIMUM BENEFIT FROM FILAMENT WINDING

If filament wound components are to achieve very high specific strength and stiffness, it is crucial to ensure that the fibre orientation lies close to the direction of principal stress. Experimental work at Nottingham (1) has shown that components wound with layers at $0^{o} \pm \Phi$ angle from the principal stress direction are particularly strong, and have good fatigue properties. The angle is best kept as small as possible as there is a rapid exponential decrease in strength as this angle increases as shown in figure 3. Fatigue life degenerates in a similar logarithmic fashion. Results to date have shown the balanced configuration (with two shear planes) in general exhibits higher strength and less scatter than a similar unbalanced laminate. Figure 4 compares balanced and unbalanced winding for the case of four layers. Further work is currently in progress to quantify the effect of this parameter on strength and fatigue properties. It is clear that any form of stress which acts primarily on the matrix in which the fibres are embedded must be kept to a minimum. It is important to minimise the void content in the material to avoid local stress concentrations from which cracks are likely to propagate.

Where a component is subject to complex loading conditions it is necessary to place fibres at many different angles, and very high specific strength will not be attainable.

GENERATING A FIBRE PATH ON THE SURFACE

In the absence of friction the fibre must follow a geodesic path or it will slip out of position. A geodesic path is simply the shortest distance between two points measured on the surface, corresponding to a great circle on a sphere and a helix on a cylinder. For practical winding it is necessary to find repeating, or closed geodesic paths which can be indexed to give full coverage of the mandrel.   Finding such a path is not trivial even for a dome-ended cylinder.   The method described by Dunbar (2) is probably the best approach for simple axisymmetric shapes.   The method will be illustrated by considering the payout eye path for winding a dome ended cylinder.

For the dome ended cylinder shown in figure 5:

The mandrel will rotate half a revolution for each of the dome ends.

The mandrel rotates $L\tan\alpha/2\pi r$ revolutions as the fibre passes along the cylinder.   For one complete circuit the fibre must pass around both dome ends,   and pass twice along the cylinder with winding angles of $+/-\alpha$.   The total number of revolutions for this process is $L\tan\alpha/r\pi+1$.

If the path taken is to be a repeating geodesic it is clear that the number of rotations must be a simple integer.   Taking 1 to be 1 m and r to be 0.2 m it is found that repeating geodesics would result from winding angles of 0,   32.14,   51.49,   62.05,   68.30 ... degrees.   These geodesics would repeat in a single pass around the ends of the cylinder.   If the desired winding angle were say 42 degrees, none of these would be suitable, and it would be necessary to look at geodesics which repeated after 2, 3 or more passes.

In order to achieve complete coverage,   the next pass should lie adjacent to,   rather than   on top of,   the first.   This can be achieved by rotating the mandrel an extra $\delta\theta$ on all movements such that $\delta\theta = w/\cos\alpha$ where w is the width of the fibre bundle or tape. For best results $\delta\theta$ should be rounded down to the nearest simple factor of 360 degrees.   The slight departure from a geodesic path implied by this indexing process does not cause problems in practice.

The movement of the payout eye can be derived from the fibre path by making a further assumption, for instance that the tow length is constant   or   that the   payout   eye   lies   in   a   known   plane   or cylindrical surface.   The payout eye position at any time lies along the tangent from the position on the surface.   A coordinate transformation from mandrel-based to machine-based coordinates is necessary for the generation of the part program.

After   the   first   complete layer has been wound,   the   effective radius   of   the cylinder will have increased,   and a   consequential small change in the winding angle must be made to ensure   repeating paths in the new layer.

This method can be extended to conical surfaces, and therefore to any axisymmetric shape,   as such a shape may be approximated to any desired accuracy by a system of cones.   The method cannot easily be extended to non-axisymmetric shapes,   and is unsuitable in any case for shapes for which no simple mathematical description exists.

An alternative way of generating geodesic paths for mandrels with simple curvature is to generate a surface development.   Geodesics become straight lines on the development,   but must pass through a

common start and end point at identical angles in order to generate
a closed, repeating loop. Where the mandrel has a plane of
symmetry it is possible to cross the two lines of intersection of
this plane with the mandrel surface at right angles. A geodesic
path with this characteristic may simply be reflected in the plane
of symmetry to give the desirable closed repeating path.

Yet another approach is to numerically solve the equation for a
geodesic line on a specified surface as described in (3). This
approach requires the use of substantial computer facilities and is
limited to surfaces which are amenable to mathematical description.

At Nottingham a surface modelling program has been developed
which allows complex surface geometries to be approximated by a
mesh of triangular plane patches. The accuracy of the surface
definition can be improved to any desired degree by refining the
mesh. Any non-branched mandrel may be described, where a non-
branched geometry is defined as one for which there is always a
single closed outline for any section taken perpendicular to the
mandrel axis. It should be noted that a simple pipe 90 degree
T junction is consistent with this definition if the mandrel axis
is parallel to the centre line of either pipe in the junction.
Concave surfaces, and "banana-like" structures are not excluded.
The capability of describing very complex, branched shapes is also
available in the form of a proprietary surface modelling program
(4). The geometrical data from this is used to generate a
triangular mesh which is entirely compatible with that generated by
our own program.

The calculation of a geodesic path over a series of plane patches
is a straightforward numerical procedure in comparison with the
solution of the geodesic equation, and can be carried out on
inexpensive computers. The IBM/XT with extended graphics and
floating point co-processor is quite suitable, with graphics
workstations such as the Whitechapel MG1 offering improved
performance at a higher but still reasonable cost. Paths which
deviate from true geodesics according to specified criteria, and
paths at a constant winding angle can also be calculated
efficiently.

The majority of the work to date has concentrated on the
modelling of axisymmetric shapes, with repeating geodesics
calculated by the reflection method. A typical example is the cone
and hemisphere shown in figure 6. A post processor is used to
transform the filament path to payout eye motion and mandrel
rotation in machine coordinates. A part program is generated which
can be down loaded to the winding machine.

THE IMPORTANCE OF USING COMPUTER AIDED DESIGN AND MANUFACTURE

Patterns can be designed and wound on the machine with a small
fraction of the time and effort required to generate and test part
programs by non-automatic methods. It is possible to explore a
variety of winding patterns at a terminal without material wastage,
and to identify at the design stage any shortcomings thereby
avoiding the costs of perhaps scrapping a component at an advanced
stage of manufacture. For complex shapes the only practical
alternative to CAD is an expensive and time consuming trial wind

procedure using a "teach-in" facility which enables the machine to remember and repeat manually input paths.

The software developed allows a geodesic to be generated from an initial point and specified winding angle, and to generate closed loops where a plane of symmetry exists. The geodesic may deviate from a true path within the constraints of a coefficient of friction which must be realistic for the actual winding environment. The possibility of finding a wider variety of closed paths, and extending the method to geometries with no plane of symmetry by an automatic surface development method, and by an iterative search procedure are under investigation. It is likely that these processes will be carried out off-line, and the user then offered a choice of pre-calculated paths in a subsequent interactive design procedure. The CAD procedure will also allow arbitrary paths to be specified as it is hoped that current work at Nottingham will remove the geodesic constraint.

NON-GEODESIC WINDING

The problem of finding closed geodesic paths which align fibres in the direction of the principal stresses is often intractable for complex geometries. The region in the neighbourhood of the ends of the mandrel usually presents the most difficulty. The fibre can be turned round using pins in many cases, and many otherwise impossible paths can be wound using carefully placed pins, or by stapling or glueing the fibre in place. Grooved mandrels offer promise for winding shapes which do not require complete coverage of the surface. It is always possible to identify local geodesic paths which lie in the desired directions, but these rarely exhibit the desired "loop back" properties. A cooperating robot will be used in future work at Nottingham to anchor the start and end points of such geodesics with adhesive or staple gun, thereby allowing piecewise continuous geodesic loops to be used with anchors at the discontinuities. A robot would also be useful for placing pre-formed shapes on the mandrel. It is likely that a future generation of winding machine would incorporate the ability to anchor fibres using a built-in attachment in the region of the payout eye. Stapling has many attractions when using wooden or rigid foam mandrels.

The use of friction to deviate from a true geodesic path depends critically on the winding environment, as the coefficient is a complex function of mandrel and filament materials and resin. When winding wet the resin acts as a lubricant, and friction is very low. Pre-impregnated tape in contrast is tacky and gives a higher coefficient of friction than dry fibres. The use of impact adhesive on both the mandrel surface and the fibres allows almost complete freedom in the choice of path. The possibility of winding almost ideal structures as a stiff preformed shape using adhesive, and then impregnating the resin at a later stage by a process such as resin injection is superficially very attractive. The chemical and structural effects of the adhesive on the resin matrix are unknown at present, and may prove major drawbacks to this process. A number of adhesives are under investigation at present. The use of a hot melt glue dispensed from a heated bath close to the payout eye may be effective, especially where the mandrel is made from a material of high thermal conductivity. Spraying an activator onto the fibre, and the use of ultra-violet and infra-red curing resins are further possibilities.

MACHINE CONTROL REQUIREMENTS FOR FILAMENT WINDING

The control requirements for a filament winding machine differ greatly from those of a traditional machine tool. The speeds and inertias are much greater for filament winding, and the cutting loads very much less. Accuracy is specified in units of 0.1 mm rather than 0.005 mm as might be the case for a precision milling machine. A further problem is that accelerations of the mandrel and fibre must be carefully controlled, and smooth high speed interpolation between several axes is essential. Ideally the fibre should be dispensed at a steady speed in order that a homogeneous matrix results, and to avoid the filament jumping out of place, or even breaking. It is unfortunate that most current purpose-built filament winding machines incorporate standard machine tool controllers. These subject the moving parts, mandrel and fibre to fierce transient accelerations and are often incapable of the required smooth multi-axis interpolation. These extremes of acceleration have caused actual damage to a mandrel constructed by glueing and sanding plywood blocks.

The large variations in mandrel inertia lead to positioning inaccuracies in non-adaptive controllers, and the available G codes are of little or no value for filament winding. At Nottingham we are constructing a special machine control system which will overcome these limitations. This will have the following features:

1 Adaptive control which will compensate for mandrel inertia.

2 Smooth interpolation on five axes.

3 Scaling of the machine speed in inverse proportion to the instantaneous radius of winding.

4 Built in functions for a range of situations which are commonly encountered in filament winding (eg cylinders, dome ends, cones etc).

5 A "pause and teach" facility to allow standard patterns to be patched together by manually specified movements.

6 A mandrel measurement mode. The controller will allow the machine to be used to inspect the mandrel geometry. A position transducer will be mounted in place of the payout eye, and the profile of a series of sections at different axial positions measured. The machine will use this information to ensure that the payout eye is not traversed into the mandrel, and to establish an instantaneous centre of winding.

7 An interlock facility will allow a cooperating robot, or other device to take action before winding proceeds. This facility will communicate the current position in space of the contact point of the filament with mandrel surface.

CONCLUSIONS

1 The advent of CNC filament winding machines has made this technique cost effective for the manufacture of composite structures of high specific strength and complex geometry.

2 The use of CAD/CAM is essential if this process is to be cost effective.

3 A new generation of machine control systems is needed to realise the full potential of the process for high speed manufacture.

4 Further work on techniques for non-geodesic winding is urgently needed.

REFERENCES

1 Owen MJ, Middleton V, Weatherby N, 'Fatigue properties of filament wound FRP materials - Progress Report No. 4', University of Nottingham Department of Mechanical Engineering Research Report, March 1986.

2 Dunbar GR, 'Filament winding', ASME Report 66-MD-86.

3 Eckold GC, Wells GM, 'Computer aided design and manufacture of advanced composites', Proc. 2nd Conference on Materials Engineering, 5-7 Nov 1985, IMechE and PRI.

4 'Duct manual', DeltaCam Ltd, Argyle St, Birmingham.

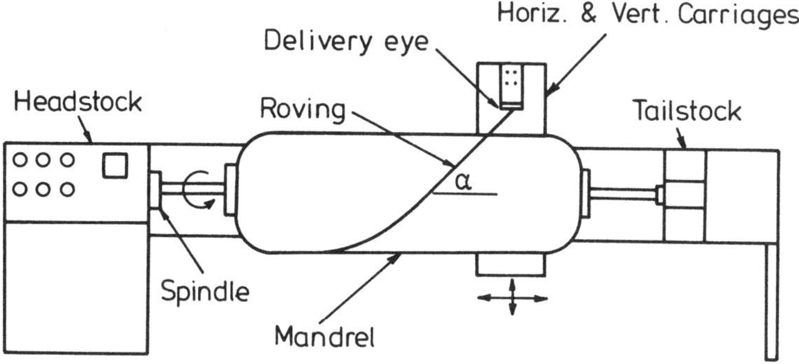

Figure 1   Typical Filament Winding Machine

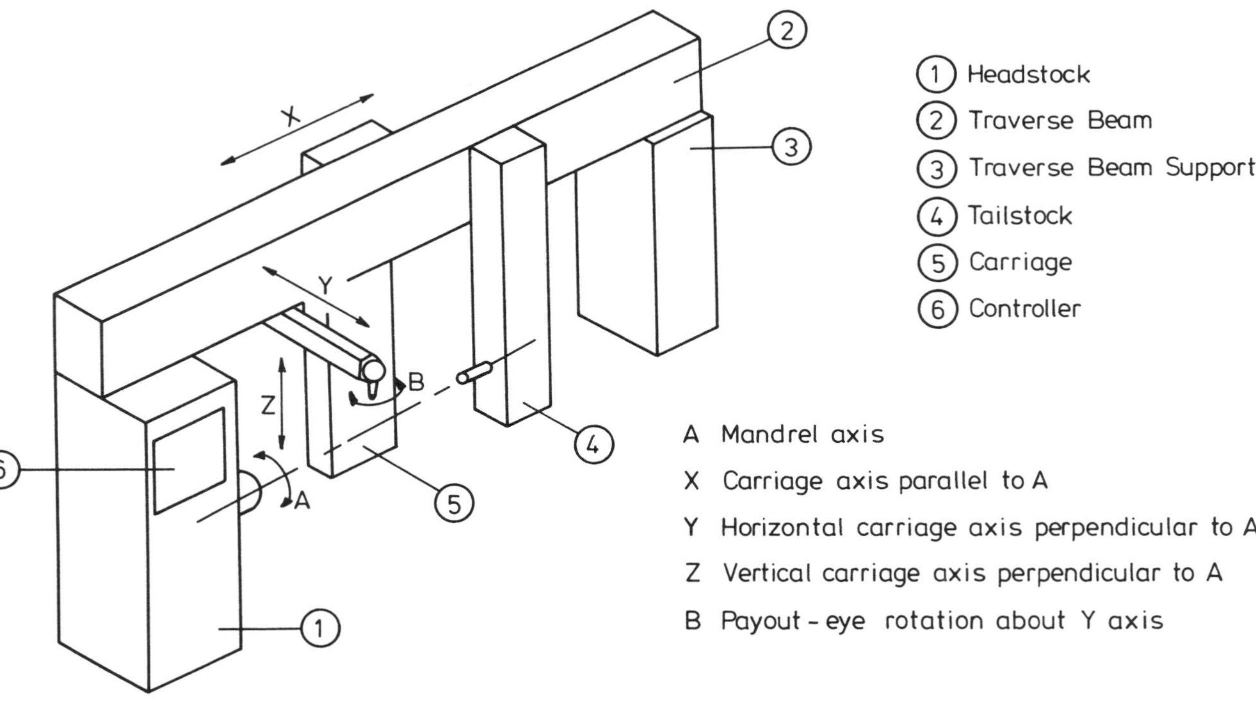

Figure 2   The Nottingham Filament Winding Machine

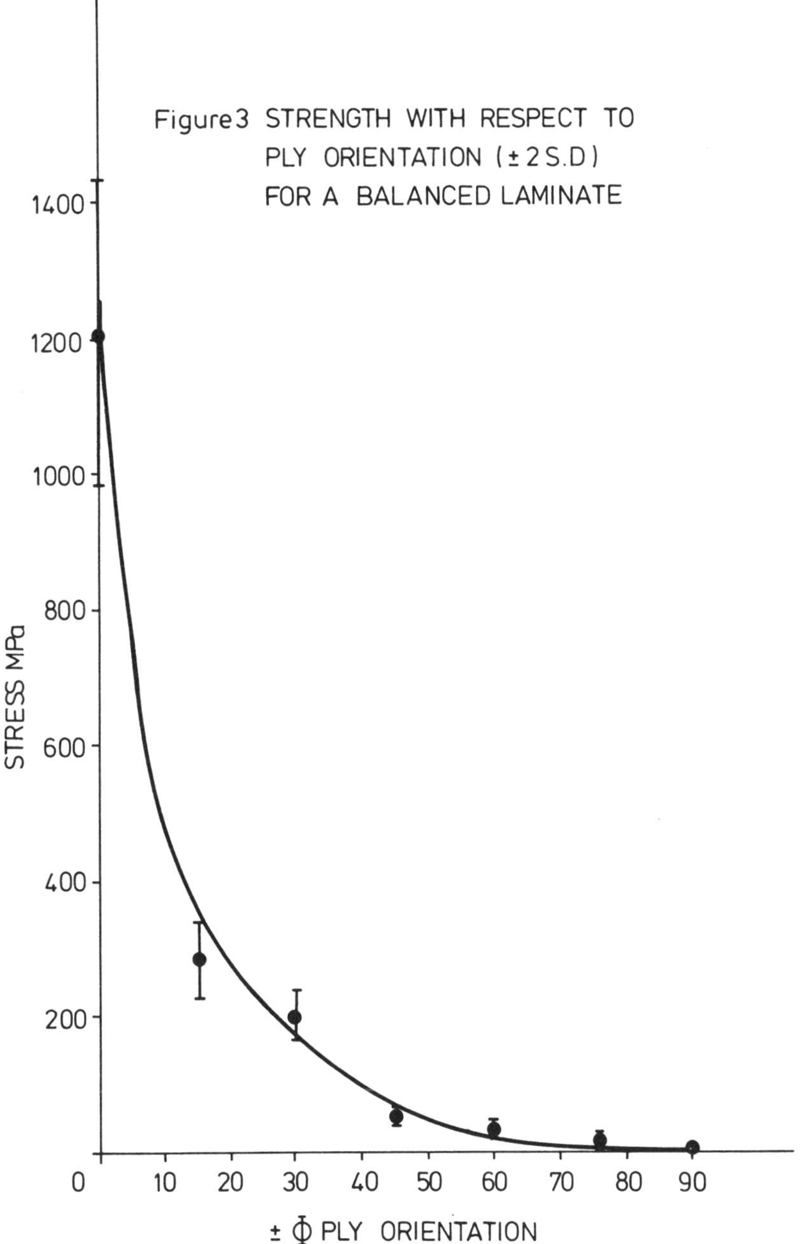

Figure 3 STRENGTH WITH RESPECT TO
PLY ORIENTATION (±2 S.D)
FOR A BALANCED LAMINATE
1400
1200
1000
800
600
400
200
STRESS MPa
0   10   20   30   40   50   60   70   80   90
± Φ PLY ORIENTATION

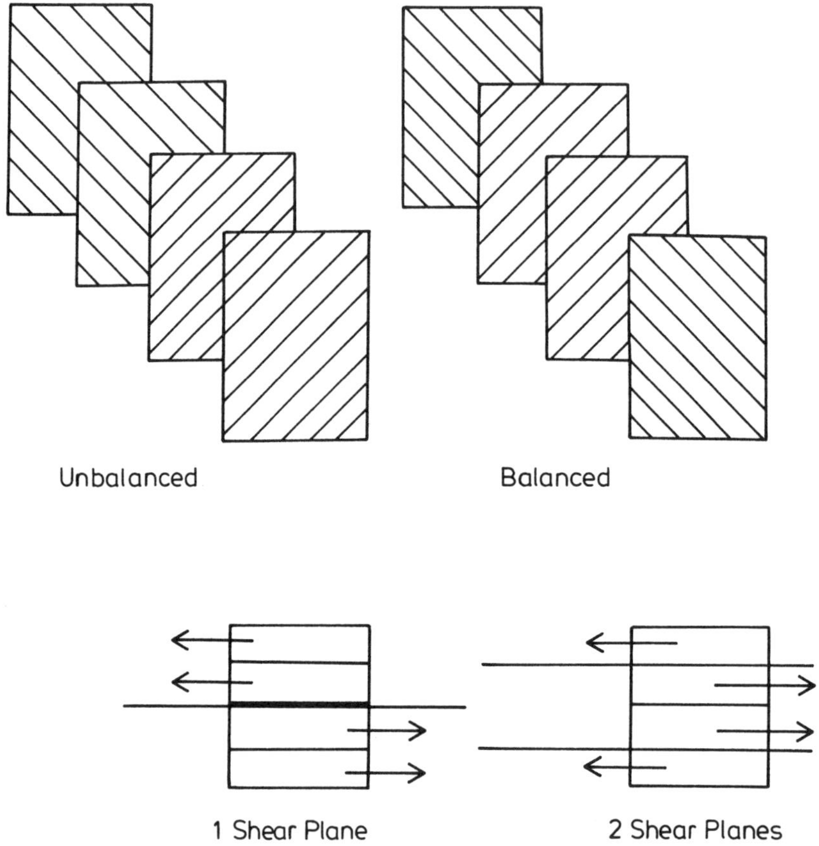

Figure 4

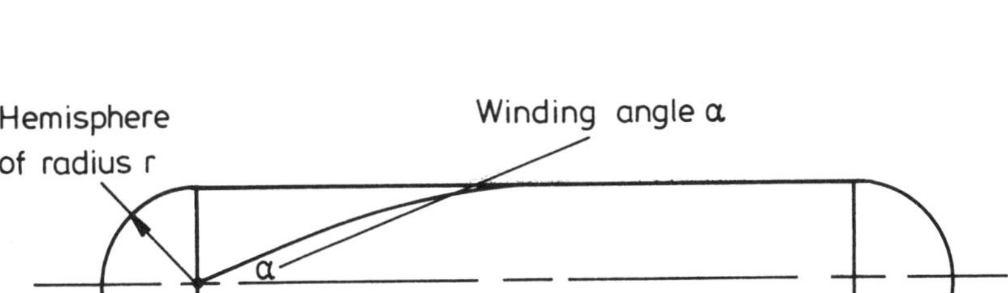

Figure 5 <u>A Dome Ender Cylinder</u>

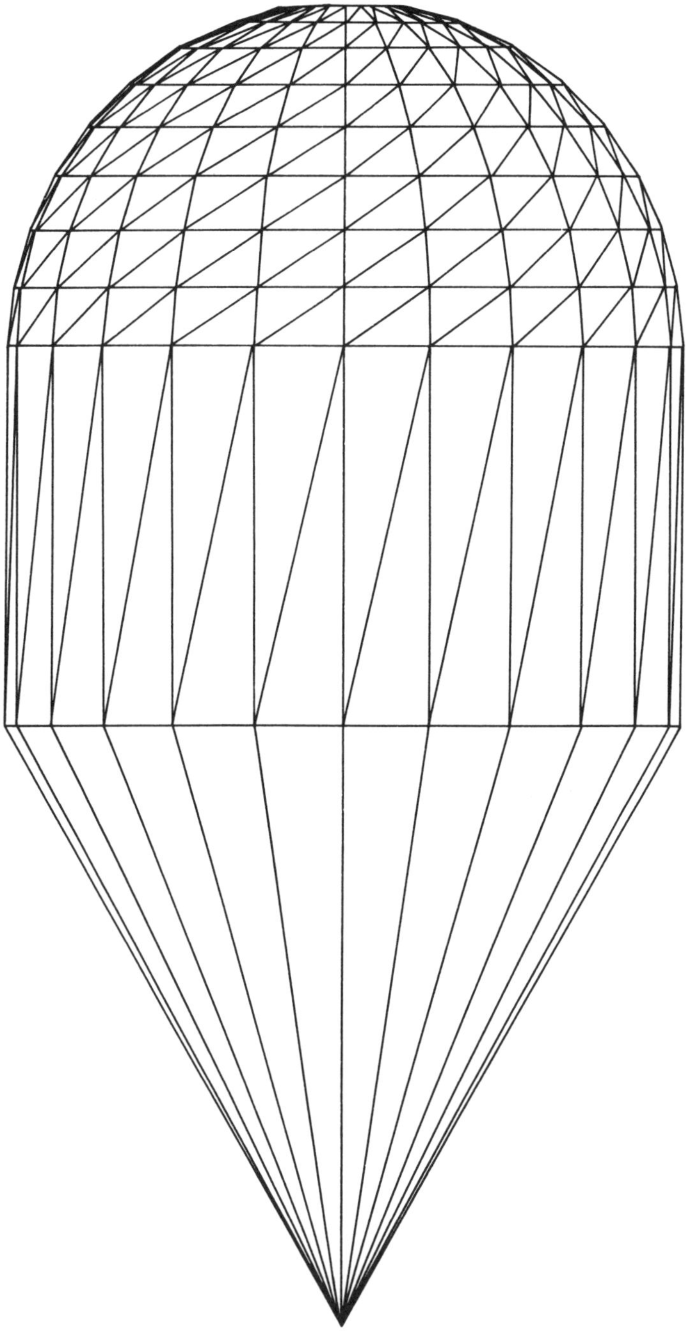

Figure 6   A Typical Surface Model

THE DEMANDS OF MODERN RUBBER EXTRUSION SYSTEMS WITH SPECIAL REFERENCE TO MULTIPLE EXTRUSION LINES AND MICROPROCESSOR CONTROLLED APPLICATIONS

M.I. Iddon*

<u>INTRODUCTION</u>

Around 1880, Iddon Brothers were amongst the first to design and manufacture processing plant specifically aimed at the requirements of the rapidly developing rubber processing industry. In 1985 we were the first company to introduce microprocessor control into the rubber process extrusion to actually control the final form of the end product. This process is universally known as the 'Computahose' and is used for extruding virtually any shape of preformed hose and tubing associated with the automobile and domestic appliance industry for example.

However, before examining this extrusion system in more detail, it is interesting to reflect back to the 1880's and research the progress, and even more interesting the time-scale, and case history of our extrusion industries evolution.

<u>SLIDE 1</u> Originally called a Forcer; this slide shows a typical extruder crica 1910/1915. Since its origination thirty years previously, little had changed, except that the belt drive from the overhead line shaft had been replaced by a self contained fixed speed electric motor.

<u>SLIDE 2</u> Around 1924 - Roller Extruders which incorporated two vertical rolls into which hot rubber was fed, and in turn, forced through a die box. The forerunner of the roller die system, and enjoyed a limited success.

<u>SLIDE 3</u> Circa 1950. Thirty years later, industry was still adopting the same basic unit. The three speed gearbox was stronger, steels were of a better quality and more able to withstand abrasion and general wear and tear, and around 15% of extruders were fitted with variable speed drive motors, which really was the biggest step forward over the seventy years in question.

 * Iddon Brothers Ltd

In the meantime, gains in technology were achieved in the materials and processes developed by industry. Synthetic rubbers have now been with us for around fifty years, and a general move was made around the mid 50's in the direction of the possibility of processing cold rubber directly into the extruder. Up to this date, all extruders being fed with hot strip.

Very briefly, and purely to set the scene, the first significant change in processing technology was made around the early to mid 1950's. The cold feed extruder appeared 70/75 years after the birth of the Extruder/Forcer. A period which covered the best part of a century of relaxed and minimal progress in equipment technology. These early cold feed extruders had, in general, a very disappointing performance and almost died a natural death. Fortunately, with the advent of continuous curing in the 1960's, serious and urgent consideration was given to eliminating these shortcomings.

SLIDE 4 Currently essential features of an efficient, reliable and versatile extruder incorporates into its design, the follow -

1.   a reliable and accurate variable speed drive, now universally D.C. with refined monitoring systems to ensure that output speeds exactly match the process requirement.

2.   a good feed roll design, coupled with a carefully developed feed pocket profile.

3.   some form, of which many are available, of controlling the temperature of each of the zone of the extruder, plus, and of equal relevance, the temperature of the scroll. Machines nowadays have up to 7 or 8 zones, and you will note, on page 9 how each polymer requires its own temperature profile for optimum performance.

4.   a range of scrolls, both vented and non-vented, to suit the process application. Here the rubber industry is more fortunate than the plastics industry, by virtue of the fact that it is possible to design a general purpose scroll which will efficiently process a wide range of natural and synthetic polymers. The differing characteristics of these polymers being largely balanced out by the recommended temperature profile.

Now we arrive at the most vital development of the whole extrusion unit, namely a predictable and versatile scroll design. Here I would stress that there are several excellent designs available, and of equal interest is the fact that broadly speaking each major design incorporates very differing approaches to the problem of producing high outputs - coupled with excellent quality and uniformity. By and large, our rubber extrusions industry has never looked kindly upon a range of differing scrolls for various product ranges and applications, unlike our related colleagues in the plastics industry, the whole gamut of output is expected from one design of scroll profile, with very little give and take.

SLIDE 5 Typically shows our approach which has become a world wide accepted standard of Extruder Scroll Design. This slide shows our H.I.M. (High Intensity Mixing) Scroll concept.

SLIDE 6 This slide shows the theory of the patented H.I.M. Scroll transcribed into actual practice.

<u>SLIDE 7</u>   This slide demonstrates what actually happens to the polymer as it progresses down the flights of -

a)   a conventional, 2 start scroll with a compression ratio of say, around 1.8/1, typically used in the cold feed extruders of the 1960's and even the 1970's vis-a-vis.

b)   the flow path of a modern scroll incorporating our H.I.M. concept.

We have now reached a stage where extrusion plant is not only reliable but totally, and I stress totally, predictable.

<u>SLIDE 8</u> (Output characteristics). This slide demonstrates what we have achieved and why we can now enter into a complete new world of microprocessor control systems., Ideally, now we have achieved the situation where the output performance of the extruder exactly relates to the scroll speed, as we can see from this graph. If the speed of the scroll is increased 50%, then the gain in output should balance at 50%.

The modern extruder, therefore, is totally predictable. The output for any profile, in any polymer or cross blend can, from theory, be exactly calculated for costing purposes, and reproduced on the production floor by an operative. The mystique enjoyed by the 'old hands' on the factory floor to manufacture the product has gone. Control systems and a general understanding of good designs have taken over.

The advancement in know-how, and improvement in processing plant over the past ten years at least equals, and very possibly exceeds that of the previous ninety years.

In many respects, the next ten years will probably see little further progress in the development of the basic extruder. However, new machining techniques, etc., will advance, hand-in-hand with realistic value engineering and help to stabilise manufacturing costs.

The title of my Paper, provides a clue as to what we can expect over the next decade. Most extruders operate on average at around 50/60% of their full potential, due to many reasons, some of which are listed as follows:

1.   the ability of the operative or ancillary processing plant to handle a higher volume of product

2.   heat input into the material being processed during extrusion, coupled with demands from industry to limit the maximum permissable termperature of extrudate to, often as low as 100°, or even less

3.   with a conventional type die head, a maximum line speed of around 30 metres per minute is rarely exceeded, limited to the frictional head build-up, etc., in the compound as it passes over the die head components. The future will see considerable development in design to overcome this limitation.

The scope of the extruder, size for size, over the last decade has increased dramatically. Typically the processor can obtain the same output from for example, a modern 90mm that he would otherwise generally have anticipated from a 115mm unit. <u>SLIDE 9</u>

MICROPROCESSORS (M.P's) AND THEIR APPLICATION

Our industries have to move into a new era of technology if progress is to be maintained. This technology must be aimed at -

1) Improving quality of production.

2) Reducing scrap levels.

3) Releasing manpower, which can be more effectively utilised in other areas.

4) A constant monitoring and correction capability to ensure that any variations in basic materials, factory services, etc., etc., are detected and their effects on the end product either totally eliminated or reduced to acceptable levels.

5) Reducing unit costs.

To achieve this, excellent progress has been made in developing, what possibly only a few years ago presented expensive, bulky computors, quite beyond the aspirations of the majority of industry to utilise either on a cost effective or reliability basis. The modern M.P.'s package is robust, compact and in real terms, low cost, coupled with its now accepted reliability we have a technlogical package that can be put to good use; a modern M.P's an exceptionally versatile tool.

We have watched M.P's become a way of life, as you can see from

SLIDE 10 45mm PVC Extruder

It is a simple matter to install a M.P. with the capability of controlling 8-12 temperature zones, and in addition, monitor and control -

1) scroll speed

2) drive motor power

3) melt pressure along the barrel

4) die pressures, et., etc.

Such systems are relatively cheap, considerably less than say five years ago, reliable and easily maintained, but by and large, customer demand is still relatively low. Perhaps over the next decade, the situation will change, only time will tell. Iddon Brothers have, in keeping with many other machinery suppliers, been endeavouring to establish the exact role of the M.P. in the field of extrusion.

We are of the firm belief that such low key demands are an insult to the intelligence of M.P's and that the efficiency of the modern extruder, engaged upon a typical range of process applications, is such that little or no benefit can be gained by the introduction of an M.P. whose sole purpose is primarily to carry out the above functions, unless of course, statistical feed back is required for reference purposes, as is the case, for example, in research and development. A more demanding and definitive role must exist for the prime requirement, and if necessary, the above functions can then all be incorporated as the secondary requirement for the M.P.

We are definitely of the opinion that the role of the microprocessor in extrusion is to be able to exactly monitor all aspects of the product being manufactured by the extruder. Currently, we have a serious vacuum in the priority development of commercially acceptable measuring systems for every day applications. However, rapid progress is taking place in the development of these requirements which will enable, for example, a continuous running dimensional check to be constantly kept upon, say for example, a hose profile. Any change in diameter or concentricity will immediately be apparent, and the information collected from the scanner will be fed in to the M.P., resulting in the appropriate die adjustments being automatically carried out. Enormous potential exists throughout the industry for such applications which would ensure that the savings on scrap products and excessive use of material would soon pay back the initial investment costs.

SLIDE 10 Demonstrates the 'Iddon Computahose®' Extrusion system made possible by the excellence and reliability of the modern extruder coupled with the allied M.P's. Essentially it comprises three units:

1. The extruder.

2. The Computahose® Variable Geometry head unit.

3. The M.P's control centre.

The principle involved is quite straightforward as any one who has ever involved themselves in the mystery of tube extrusion will appreciate.

SLIDE 11 If an outer circular die is geometrically located exactly central with a fixed inner circular die, a tube will be produced with uniform wall thickness, i.e. a straight, true concentric tube. However, as you will note from the slide, if the outer die is not concentric, unequal flows are generated, thus producing a 'bent' tube. By thus controlling the exact location of the outer die relative to the inner die is the principle of the Iddon Computahose® Extrusion system. Straights, (S)-Bend Radii, and Bend Vector angles can be accurately produced to generate any form of 'shaped hose'. Perhaps at this stage it would be opportune to demonstrate the process by showing a video film.

This video film will, I hope, conclusively confirm my predictions. A process has been developed whereby a sequence control system that precisely locates the outer die in relation to the inner die gives the ability to at all times control the relative die positions, and enables us to automatically and very simply, develop a fully M.P. controlled extrusion system to produce a wide range of formed tubes, either as waste pipes and traps manufactured from standard compounds, e.g., Butyl or E.P.D.M. SLIDE 12, or by using Monsanto's Santoweb ®, which is a fibre reinforced material, a full range of automotive radiator and heater hoses can be continuously extruded at a rate of up to 10 units/minute SLIDE 13 and SLIDE 14.

F I L M

Those of you who, like myself, have grown up in the slide rule era, possibly have an inherent distrust of the M.P's and their technology. Nothing could be further from the truth. Our admittedly short experience in this field of activities has taught us that M.P's, used with intelligence are, without a doubt, an extremely effective processing tool. Very briefly the next two slides will demonstrate how we, employing our own skills and initiative, developed what I believe to be not only a very professional programme, but a programme that, as customer experience and feed back has proven, is one that is easily followed and almost impossible to incorrectly implement.

COMPUTAHOSE® PROGRAMME

SLIDE 15 (OPTIONS SCREEN)

This display gives six options, marked:

1) BLOCK INDEX

2) SAVE DATA

3) LOAD DATA

4) ENTER NEW DATA

5) EDIT DATA

6) RUN DIE HEAD

The required Option 1 to 6 is selected by pressing the appropriate number key.

OPTION 1   BLOCK INDEX   (press key marked [1])

This provides a quick and easy reference as to which blocks are free (marked 'NOT USED') and which Blocks are programmed (marked with your Block Code). Pressing [SPACE BAR] reverts to the Options Screen. (SPACE BAR is the long key situated at the bottom of the keyboard).

SLIDE 16
OPTION 2   SAVE DATA (press key marked [2])

This section saves data for future use, and will save-to-disc all profiles programmed into the Main Block, and filed under the code name and number.

OPTION 3   LOAD DATA (press key marked [3])

This section is used to load back into the M.P. information that has previously been saved on disc.

Place data disc in the drive. M.P. instructs

PRESS [C] TO LIST CONTENTS OF DISC

PRESS [L] TO LOAD DATA

Pressing [C] will list the complete contents of the disc.

Select the Block you wish to load, press the key marked [L]

M.P. instructs **ENTER BLOCK CODE.**   Enter this in the normal way, then

press [RETURN]. On completion, the MP will revert to the Options Screen.
COMPUTAHOSE® PROGRAMME ACTUAL

OPTION 4   ENTER NEW DATA   (press key marked [4])

This section is used to input a hose profile program. Throughout this section the M.P. raises a series of questions as the following slides will demonstrate - answered with the appropriate key presses, followed by [RETURN]

The normal steps are as follows:

a)   Enter BLOCK number

b)   Enter **BLOCK CODE NAME (maximum 10 characters)** AND Key in Product
     Code.

c)   SLIDE 17   Enter **EXTRUDATE OUTPUT SPEED (SECS PER 100mm),** By
     keying in your appropriate linear output speed.

d)   SLIDE 18   Enter **BEND STRAIGHT OR END (B/S/E)?** Press [B] to
     program a Bend, press [S] to program a Straight, press [E] if entry
     data is complete.

     PRESSING [B]   M.P. queries: **ARE YOU CORRECT**

     Press either [Y] for Yes, or [N] for No.

     [Y] will ask for the displacement value - this refers to the required
     radius of the bend measured to the centre of the hose. Key in the
     relative value to produce the desired radius.

e)   SLIDE 19 M.P. instructs **BEND VECTOR ANGLE (1-360)**
     This refers to the sector of the die where bend is required.
     360°-180° are presented vertically; 90°-270° horizontally in the
     conventional manner.

     PRESSING [S] for straight centralises the outer die and any length of
     straight section can now be keyed into the programme.

     When profile data entry is complete, press [E] to return to Options
     Screen.

     OPTION 5   EDIT DATA   (press key marked [5]

     This section allows editing of any Blocks of data stored in the Data
     Bank.   The M.P. instructs **ENTER THE BLOCK NUMBER YOU WISH TO
     EDIT.**   Key in a number from 1 to 20.   Press [RETURN] and the system
     will then display the selected programme and enable the operative to
     selectively 'EDIT' any section of the hose construction sequence.

     OPTION 6   RUN DIE HEAD   (press key marked [6]

     This section is used to run your program on the die head, and is where
     the M.P. converts the entered information into a series of electronic

signals and pulses required to operate the die head.

1) M.P. instructs **ENTER THE BLOCK NUMBER YOU WISH TO RUN**

2) Key in the appropriate coding number 1 to 20

3) The M.P. then automatically activates the air and hydraulic systems. A short delay takes place to allow the systems to attain operating pressure.

4) M.P. instructs the operative to zero, i.e. **CENTRALISE THE DIES USING THE X AND Y ADJUSTMENT CONTROLS.** These are situated on the main die head trolley. Once the dies have been centralised correctly, press [S].

5) The M.P. now runs the hose profile with automatic guillotining on completion.

6) The sample hose can be examined to establish its accuracy to the design specification.

7) If any individual section of the hose requires modification, select Option 5 - Edit and Correct. Modifications can be checked as required.

8) Once the hose form has been certified as correct enter the component quantity required and the Computahose® system will proceed to produce the quantity required.

I hope that the text enclosed with my paper does not serve to put you off M.P. applications. In actual practice we have found that potential users of the system can, with limited supervision, successfully programme their own hose profile within thirty minutes of being introduced to the equipment, and once experienced, dependant upon the complexity of the hose, can easily carry out a typical programme operation in around 2/3 minutes maximum.

Ladies and Gentlemen, we feel that we have introduced the M.P. into the role where it is both cost effective and functional. Five years ago, such a development, even if it had been possible, would inevitably have been expensive owing to the high cost of the M.P. Today, it is both practical and possible. We know for a certainty, that the application you have just seen is merely the tip of the iceberg. Rapid development in die location and control will take place, hand-in-hand with the availability of relatively low cost measuring and monitoring systems, and the M.P. will become as much a part of our life in the plastics and rubber extrusion world as the Extruder itself.

A GENERAL GUIDE TO OPERATING H.I.M. VENTED SCROLLS

This scroll is designed to give increased mixing coupled with
higher output, and differs from standard Cold feed Extrusion
scrolls by the fact that the H.I.M. scroll is normally heated
and the barrel cooled. As an approximate guide to Extruder
temperatures, we submit the following:-

E.P.D.M. - 60 Shore

Zone 1 (Feed)	Zone 2	Zone 3	Zone 4	Zone 5 (Head)	Scroll
50°C	50°C	40°C	45°C	70°C	60°C

P.V.C. NITRILE - 60/70 Shore

Zone 1 (Feed)	Zone 2	Zone 3	Zone 4	Zone 5 (Head)	Scroll
60°C	55°C	70°C	55°C	95°C	80°C

NATURAL RUBBER - 35 Shore

Zone 1 (Feed)	Zone 2	Zone 3	Zone 4	Zone 5 (Head)	Scroll
50°C	50°C	40°C	40°C	80°C	60°C

NATURAL RUBBER - S.B.R. - 60 Shore

Zone 1 (Feed)	Zone 2	Zone 3	Zone 4	Zone 5 (Head)	Scroll
60°C	60°C	55°C	50°C	90°C	75°C

NEOPRENE - GRADE T

Zone 1 (Feed)	Zone 2	Zone 3	Zone 4	Zone 5 (Head)	Scroll
50°C	55°C	45°C	65°C	70°C	90°C

NORSOREX - 23 Shore

Zone 1 (Feed)	Zone 2	Zone 3	Zone 4	Zone 5 (Head)	Scroll
50°C	40°C	40°C	40°C	75°C	60°C

It must be appreciated that these temperatures are only approximate
and as compounds vary from user to user, a certain amount of
experimenting may be necessary in order to find the perfect
conditions for your particular compounds. It is advisable to run
the scroll at 10 - 15 rpm on start-up, to avoid excessive loading
of the motor, bringing the Extruder to normal running speed once
the rubber has reached the die head.

A GENERAL GUIDE TO OPERATING H.I.M. SCROLL

This scroll is designed to give increased mixing coupled with
higher output, and differs from standard Cold feed Extrusion
scrolls by the fact that the H.I.M. scroll is normally heated
and the barrel cooled. As an approximate guide to Extruder
temperatures, we submit the following:-

NATURAL RUBBERS

Zone 1 (Feed)	Zone 2	Zone 3	Zone 4 (Head)	Scroll
50 - 60°C	40 - 50°C	40 - 50°C	80 - 90°C	80 - 90°C

S.B.R.

Zone 1 (Feed)	Zone 2	Zone 3	Zone 4 (Head)	Scroll
45 - 55°C	35 - 50°C	35 - 50°C	75 - 90°C	75 - 90°C

E.P.D.M.

Zone 1 (Feed)	Zone 2	Zone 3	Zone 4 (Head)	Scroll
45 - 55°C	35 - 50°C	35 - 50°C	75 - 90°C	75 - 90°C

BUTYL, NITRILES & NEOPRENES

Zone 1 (Feed)	Zone 2	Zone 3	Zone 4 (Head)	Scroll
45 - 55°C	35 - 50°C	35 - 50°C	65 - 75°C	65 - 75°C

It must be appreciated that these temperatures are only approximate
and as compounds vary from user to user, a certain amount of
experimenting may be necessary in order to find the perfect
conditions for your particular compounds. It is advisable to run
the scroll at 10 - 15 rpm on start-up to avoid excessive loading
of the motor, bringing the Extruder to normal running speed once
the rubber has reached the die head.

SLIDE 1   HOT FEED EXTRUDER c. 1905

SLIDE 2   ROLLER TYPE EXTRUDER c. 1920

SLIDE 3   X. L. HOT FEED

SLIDE 4   75mm W.X. Cold feed vented extruder

SLIDE 6   H.I.M. Scroll

SLIDE 9   Pipe Seal Profile

SLIDE 10   M.P's control on a PVC extruder

SLIDE 11   Geometrical layout of tubing head

SLIDE 12   Radiator hose samples

SLIDE 13
           Computahose ® plant layout
SLIDE 14

SLIDES 15 to 19   Examples of the Computahose® M.P's programme

Slide representations not supplied with this document.

DEVELOPMENTS IN REINFORCED REACTION INJECTION MOULDING FOR INCREASED PROCESSING FLEXIBILITY

P.D. Coates* and A.F. Johnson+

Reinforced reaction injection moulding (RRIM),
a low pressure reactive moulding process
offering substantial processing and
manufacturing advantages is discussed. A
unique computer controlled four stream RRIM
machine is presented as a means of increasing
the flexibility of RRIM in terms of chemical
formulation design, control of processing, and
the manipulation of product properties with
materials such as commercial polyurethanes and
interpenetrating polymer network alloys.

1. Introduction: The RRIM Process in Context.

It has been estimated that around 25% of all articles made
from polymeric materials are manufactured using reactive
processing methods, i.e. technologies which directly involve
polymerisation reactions in the production cycle (1). For
example, monomer casting, the hand lay-up of polyester resins and
the injection moulding of amino resins all rely on polymerisation
reactions during the manufacturing process to obtain good final
product properties. Fig. 1 shows schematically the diverse field
of reactive processing techniques.

The more traditional reactive moulding routes (open or
closed mould) have generally been characterised by rather long
cycle times with respect to those required for mass production,
such as in injection moulding of thermoplastics, but over the
last decade the development of one of the established reactive
processing techniques, reaction injection moulding (RIM),
which has low in-mould pressures, has provided a serious
competitor to conventional injection moulding for larger
articles.

In the RIM process, liquid reactants of low molecular weight
are rapidly mixed just prior to filling a mould, in which
polymerisation occurs to form the finished solid products. This
technique is well suited to handling polyurethane (PU) materials
and in a low pressure dispense form has long been used for the
manufacture of shoe soles and a wide range of PU foam products.
High pressure reinforced reaction injection moulding (RRIM) was
developed from PU RIM in an attempt to produce polymer composites

* School of Mechanical and Manufacturing Systems Engineering
+School of Chemistry and Chemical Technology, University of
Bradford, Bradford, W. Yorks, BD7 1DP, UK.

with improved physical properties compared with those of PU RIM polymer matrices, in commercially acceptable cycle times (2-6).

Short fibre or particulate reinforcing fillers were initially used in PU RRIM, hammer milled glass being the most popular. Other reinforcing agents include chopped strand glass, flake glass, wollastonite and mica. Such fillers are generally incorporated into one reactant (the polyol) prior to injection - but with the drawbacks of increasing the viscosity of the reactant and machine wear problems, together with, possibly, unwanted anisotropy in product properties. More recently there has been strong interest in continuous fibre mats preplaced in the mould as more efficient reinforcing agents, with a single reactant (i.e. resin injection) or the reacting mixture of two fluids (i.e. 'RIM over mats') then being injected into the closed mould.

The RRIM process has been developed predominantly for PUs, which offer a wide range of properties through formulation variation. However, RIM and RRIM of polyamides has been commercialised (7), and the processing of polyesters, epoxies, acrylics, acrymates, phenolics and interpenetrating polymer networks (IPNs) by RRIM techniques is being pursued. The manufacturing systems employed in RIM and RRIM processing are diverse (Fig. 2). These range from a single dispensing / mixing station feeding a single mould (economically justifiable for long production runs) to multiple mixhead systems feeding carousels of moulds from a central bulk reactant storage station, with various intermediate permutations.

The manufacturing technology has developed rapidly, mainly in response to the automotive industry interest in RRIM for the manufacture of car body parts, and its potential is by no means exhausted. A major reason for this lies in the low mould pressures encountered in RIM and RRIM; jet impingement occurs at injection pressures typically around 200 bar, but most of the pressure energy is dissipated in mixing, leading to in-mould pressures generally lower than 1 bar. Such low mould pressures allow the use of relatively low strength, relatively cheap materials for prototype moulds. For example, epoxy resin and sprayed metal or electroformed nickel tooling have been found suitable for low to moderate volume RIM and RRIM production useage, although tool steel remains the preferred mould material for high precision, long production runs. The mould presses required for large area parts (of the order of square metres) need only develop clamping forces of the order of tens of tonnes. Similarly, smaller moulds such as might be encountered on a carousel system can be very simply clamped, e.g. by actuated toggle bolts. A simple comparison between a RRIM dispensing machine/clamp and a thermoplastic injection moulding machine having the same injection power is given in Table 1; whilst such comparisons can be misleading due to the significant differences in the two processes, they can also indicate some of the potential advantages offered by RRIM.

In the areas of machine control and automation, RRIM equipment did not initially enjoy the same level of sophistication as injection moulding. However, commercial microprocessor based RRIM machine sequence controllers are currently available, and there is a welcome trend of a rising level of adequate instrumentation on the newer RRIM machines. Automation, much in evidence in injection moulding, is also beginning to be exploited in RRIM; there is clearly potential for automated demoulding and automated preplacement of reinforcing mats, for example.

Cycle times for RIM and RRIM tend to be somewhat longer than those for injection moulding for similar size parts, but RRIM cycle times have been quite significantly reduced over the past few years. Demoulding and mould surface preparation with release agent are the greatest contributors to overall cycle time and have consequently received attention from chemical system formulators and machine and press manufacturers. Chemical system modifications for 'self releasing' PUs have been developed in an attempt to decrease demould and mould surface preparation times; in-mould painting of PU parts are available, offering other time saving possibilities. Mould press movement times have been reduced in parallel attempts to decrease equipment 'dead time'.

Although this introductory section seeks to set RRIM processing in context, it is perhaps also salient to comment on the actual useage of reinforced RIM. A considerable difference in attitude to RRIM is observeable in North America, compared with that in Europe; in the USA for example, PU RRIM with at least 10% hammer milled glass is used commercially for car body panel production, and customer acceptance has been achieved. There the RRIM process has found a niche, and its use appears to be expanding. In Europe however, there are only a few RRIM processors, again mainly supplying the car industry, and the end-user attitude appears to be rather more conservative. Nevertheless, there is a high level of interest in RIM and RRIM processing and products in Europe, with perhaps a wider range of products being considered.

In general, the overall forward looking viewpoint has to be that reactive processing, in which the polymer and final article are made simultaneously, will represent a growing sector of polymer processing worldwide, and it is apparent that RRIM technologies are proving suitable for exploitation of the advantages of reactive processing in the context of mass production.

2. Increased Processing Flexibility in RRIM: a multiple stream RRIM machine.

The RRIM process has been represented as one offering considerable manufacturing flexibility - in addition to the design flexibility afforded by processing relatively low viscosity fluids at low mould pressures. This paper focuses upon aspects of our research, undertaken in collaboration with a consortium of companies, which seeks to add to the scope of RRIM by increasing the options for polymer design and processing using our unique multiple stream RRIM processing facilities.

2.1 The case for multiple stream RRIM

In all the diversity of RRIM fillers, chemical formulations

and manufacturing systems, dispensing of two reactant systems, typical of PUs, has been dominant, with minor third streams being added for colourants or foaming agents in some instances. However, the blending required for such a two reactant approach imposes severe limitations on the formulator and constrains the development of RRIM chemistries, since reactants may not physically mix with others necessary to the process, and there are combinations of catalysts and reaction promoters which are not suited to two reactant formulations. It would therefore be highly advantageous to have RIM or RRIM equipment capable of accurate dispensing of more than two streams (here called 'multiple stream RRIM'), to increase the options available to the formulator.

Such multiple stream RRIM equipment makes feasible the investigation of RRIM of a potentially attractive class of new materials, namely interpenetrating polymer networks (IPNs). These provide an interesting route to the toughening of brittle polymers, or the enhancement of the flexural modulus of elastomeric matrices by combining two or more different polymers in the form of an alloy (3,8-10). There are many methods of producing these polymer alloy structures. For RIM or RRIM the several distinct polymerisation processes must occur rapidly and essentially independently in the mould cavity. Binary and ternary IPNs involving PUs, polyester, vinyl, acrylic, polyamide or epoxy materials are being explored: for example PU may be used to enhance the physical properties of epoxy or acrylic resins, or vice versa. It is apparent that these complex materials can be more readily studied if the reactants can be individually dispensed.

A further attraction of accurate multiple stream dispensing is the potential of influencing product properties by altering the ratio of reactants during the course of injection. Since injection times are short - of the order of 1 second - the dispensing equipment must not only be carefully controlled but must be capable of responding to control signals extremely rapidly. In general, all RRIM equipment should be designed to have sufficiently rapid dynamic responses to cope with fast RRIM chemistries. With faster polymerisation rates, the dynamics of the chemistry becomes increasingly convolved with the dynamics of the machine; from a research viewpoint it is very important to have machines with well identified dynamics. From a production viewpoint machines are required which are accurate, repeatable machines, and capable of handling RRIM chemistries at the appropriate rates.

2.2 Multiple stream RRIM machine development

Our RRIM research programme has attempted to take account of the requirements for both the close control and flexibility in RRIM processing outlined above, in order to develop the RRIM process and novel RRIM materials. A major aspect of our philosophy has been to combine engineering and chemistry considerations in order to deconvolve the influence of RRIM machine and chemistry variables on final product properties (11). The use of computer controlled machines with carefully identified dynamics has been central to the work.

A computer controlled four stream RRIM machine was designed and built, using as a guide the proven performance of our two stream machine which has been described elsewhere (11,12). The four stream machine, shown in Fig. 3 is modular and is capable of dispensing from one to four separate chemical streams at freely programmable ratios. It was primarily designed to study RRIM of IPNs.

The machine is housed in a purpose built 'Zone 2' safety standard laboratory, to cater for flammable reactants. Stainless steel reactant storage tanks and injection cylinders are enclosed in temperature cabinets designed to operate up to 150C. Each injection unit is hydraulically driven from an accumulator and lance movement is controlled by a Vickers Systems Ltd. KG series electrohydraulic servo valve; this provides very rapid, essentially load independent flow control. Each KG valve receives its control signal voltage from a DEC PDP 11/23 computer; linear displacement transducers sense injection lance position for closed loop feedback control and sequencing. The dynamic response of each injection unit has been carefully identified (11,12).

The real time control and monitoring cycle time is approximately 10 msec for four stream operation. Graphical output of machine and process variables monitored in each cycle and numeric assessments are available immediately after a shot. Injection rates up to 3.5 litres/sec are attainable; the machine is also capable of operating with a variety of mixheads through the freely programmable closed loop control available. It has been designed to allow ready changes of hardware, e.g. injection lance/cylinder combinations to alter the volumetric range of operation, and software to accomodate any sequencing and control modifications required. Fuller specifications of the machine have been reported elsewhere (13,14).

The machine is complemented by a range of instrumented moulds with close temperature control, and a commercial hydraulic press (ESU Ltd). The mould is essentially a batch reactor, and mould temperature control is important since it influences both the reaction rates and product morphology.

2.3 Multiple stream RRIM programmes

Two aspects of the work being undertaken in our laboratories, reported in refs. (14,15), will be discussed here, in the context of further increasing the flexibility of RRIM processing. The first represents a novel method for processing PUs, illustrating the potential of multiple stream RRIM for potential manipulation of product properties. The second covers the production of a PU- acrylic IPN using RIM technology, in a commercial timescale.

(i) Novel PU processing:

A novel technique for the manufacture of PUs is illustrated by the results shown in Fig. 4. Three stream injection was employed, using a Krauss Maffei KM16-4KF high flow rate recirculation mixhead. One stream was a modified liquid isocyanate, the second contained a low molecular weight chain extender and the third contained a high molecular weight polyether polyol. The catalysts, dibutyltindilaurate and an amine, were distributed in the diol stream.

This technique offers great processing and production flexibility. For example, on a shot by shot basis, the ratio of polyol and chain extender can be varied to produce PUs with different hard block content, and hence different flexural modulus and thermal characteristics. Also, a fourth stream might contain polyol with reinforcing agent, so providing even greater choice of final product properties through freely programmable stream ratios. Furthermore, because of the rapid response of our injection units in closed loop control, it is possible to vary the ratios of the reactants throughtout the shot in order to produce variations in product properties throughout the moulded article.

This latter option is commercially available for two component PU foam casting in a basic form, for the manufacture of dual hardness seat cushions, with a switchover from one foam to the second occuring in around 400 msec. The option being presented here easily embraces the commercial example and extends it to multiple stream mode, with closed or open moulds, and an estimated switchover time to the new steady state ratios of under 50 msec. With such rapid changes in ratio being feasible through the fast responding KG valves, in the case of open mould pouring the mixhead movement would be rate limiting. In the case of closed mould injection with a fixed mixhead it is clearly important to have some knowledge of mould filling patterns in order to obtain the desired property distribution in the moulding. Work is proceeding in this area.

(ii) RIM of PU-Acrylic Interpenetrating Polymer Network Materials:

There have been numerous reports in the last decade of bench syntheses of IPN alloys, the bulk of the work involving reaction rates considerably slower than those which would be acceptable for RRIM production cycle times (3,8-10)

We have examined the formation of IPNs of PU and acrylic polymers for RIM and RRIM, and have successfully fabricated plaques (approximate dimensions 900 x 290 mm, of thickness 3 to 6 mm) using the multiple stream machine, in cycle times of approximately two minutes. These RIM IPNs have been based on commercially available reagents, and comprise a polyetherurethane chain extended with butane diol, and a cross-linked polymethylmethacrylate. This appears to be the first successful moulding of an IPN material using RRIM technology, operated under essentially commercial production conditions.

The major problem of achieving acceptable RRIM cycle times with IPNs involves the judicious selection of catalysts and mould temperatures whilst achieving acceptable product properties. We have observed synergistic effects and enhanced overall reaction rates for some of the catalyst/initiator combinations required for the two polymerisations, but with others quite the reverse has been observed. Achieving complete conversion of the vinyl monomer in the cycle times of interest proved to be difficult. Even though the overall polymerisation process is complete in less than approximately 30 seconds and a truly simultaneous interpenetrating polymer network (SIN) might be expected to have been formed, it has been our experience that the PU is formed well before the acrylic system has reached high conversion. This has considerable significance in relation to product morphology, and hence final product properties.

The correlation between machine operating conditions and resultant product morphology is far from simple, and we are currently engaged in the full characterisation of the process and products. Clearly, a route to commercial production of an attractive class of new materials is being opened up.

3. Concluding Comments

Reactive processing techniques, and in particular RRIM, are emerging as competitive manufacturing methods even to traditional mass production routes such as injection moulding, offering energy efficiency, manufacturing systems flexibility and a growing range of suitable polymerisations and reinforcing fillers.

The work discussed here, based on continuing research programmes in our laboratories using computer controlled RRIM machines, has indicated two routes to further increasing the manufacturing flexibility of the RRIM process, through computer controlled multiple stream RRIM. The first route involved a novel method of producing PUs which would enable shot by shot variation of product properties, or variation of product properties with position in a moulding. The second route has lead to the first successful moulding of an IPN material, based on commercially available precursors for PUs and acrylics, using RRIM technology. Such progress has been enhanced by close collaboration at the chemistry - engineering interface, in our view a vital interface to explore if RRIM is to develop its potential.

Acknowledgements

The RRIM research programme has been supported by the Polymer Engineering Directorate of SERC, and a consortium of companies including BL Technology Ltd, Holden Hydroman Ltd, ICI (Organic and Mond Divisions) plc, Lucas plc, Pilkington Bros-Fibreglass plc, Scott Bader Ltd and Vickers Systems Ltd.

References

1. Lee L J, _Rubb. Chem. Technol._ _53_ (1980) 542
2. Becker W E (Ed.), 'Reaction Injection Moulding', Van Nostrand Reinhold, New York (1979)
3. Coates P D and Johnson A F, _Plast. Rubb. Process Applicn._ _1_ (1981) 223-238
4. Johnson A F and Coates P D, _Shell Polymers_ _7_ (1983) 75-79

5. Woods G, 'Flexible Polyurethane Foams: Chemistry and Technology', Applied Science, London (1982)

6. Kresta J E (Ed.), ' Reaction Injection Moulding and Fast Polymerisation Reactions', Plenum Press, New York (1982)

7. Skirha M D, _ACS Poly. Preprints_, _25_ No. 2 (1984) 298

8. Frisch K C, Klempner D and Frisch H L, _Poly. Eng. Sci._, _22_, (1982) 1143

9. Pernice R, Frisch K C and Navare R, _J. Cell. Plast._, (March/April 1982) 121-127

10. Hsu J T and Lee   L J, _Proc. 3rd Int. Conf. React. Process.Poly._, Strasbourg (1984) 67-74

11. Coates P D, Sivakumar A I and Johnson A F, _Poly. Process Eng._, _3_, Nos 1 and 2 (1985) 159-171

12. Coates P D, Sivakumar A I and Johnson A F, Proc. _3rd Int. Conf. React. Proc. Poly._, Strasbourg (1984) 95-106; to be published in _Poly. Process. Eng._ Vol. 4

13. Coates P D, Sivakumar A I, Johnson A F and Armitage P D, _Proc. Polym. Proc. Mach._, PRI London (1985), 13/1-9

14. Coates P D and Johnson A F, _Proc. Utech '86_ , Crain Communications (1986)

15. Coates P D, Johnson A F, Armitage P D, Hynds J and Leadbitter J, Polymer Processing Society 2nd Ann. Meet., Montreal (1986); to be published in Polym. Eng. Sci.

16. Coates P D, _Proc. Polycon '83 'Reactive Processing'_, 55-69 PRI, London (1983)

Table 1: Comparison of commercial injection moulding and RRIM machine capabilities of equivalent injection power (after ref. 16).

	Injection Moulding	RRIM
Max. shot (litres)	1.47	251 in 5 sec
Max. injection rate (l/s)	0.51	3
Installed injection power (kW)	45 <------------------>	45
Inj. energy/ kg shot (kJ/kg)	89	18
Total installed power (kW)	69	75
Clamp force (kN)	2200	250
Max. Cavity pressure (bar)	1400	3
Max. cavity projected area (sq. m)	0.016	0.8
Clamp speed (mm/s)	762	130
Total m/c weight (t)	13	9
Clamp weight/cav. proj. area (t / sq. m)	400	7
M/c footprint / cav. proj. area (sq. m / sq. m)	320	20

'SINGLE REACTANT' SYSTEMS

   <u>Cyclic</u>:

   Injection Moulding
   Compression Moulding
   Adhesive Dispensing
   Casting
   Transfer Moulding
   Vacuum Impregnation
   Resin Injection
   Reactive Rotational Moulding

   <u>Continuous</u>:

   Extrusion
   Pultrusion

'MULTI-REACTANT' SYSTEMS

   <u>Cyclic</u>:

   Hand lay-up
   Closed mould casting
   Adhesive Dispensing
   RIM / RRIM

   <u>Continuous</u>:

   Extrusion
   Casting

<u>FIG. 1</u>   A   simplified   spectrum   of   Reactive   Processing
Techniques

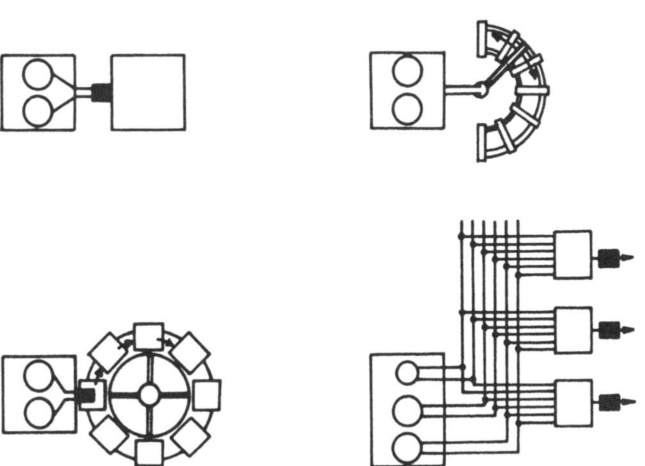

<u>Fig. 2</u> Schematic diagrams of a selection of RIM / RRIM
      manufacturing systems configurations

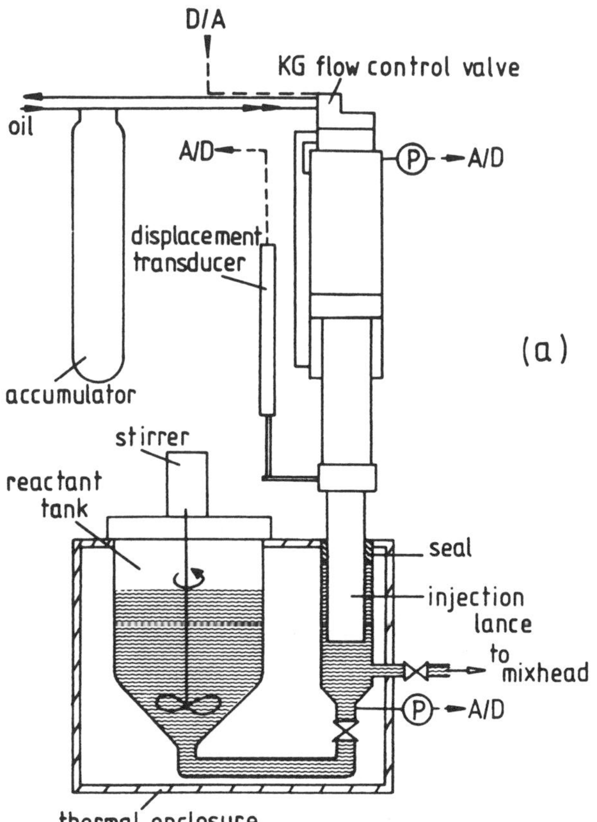

Fig. 3 University of Bradford 4 stream RRIM machine: (a) schematic of a modular injection unit; (b) the equipment installed in our 'Zone 2' laboratory

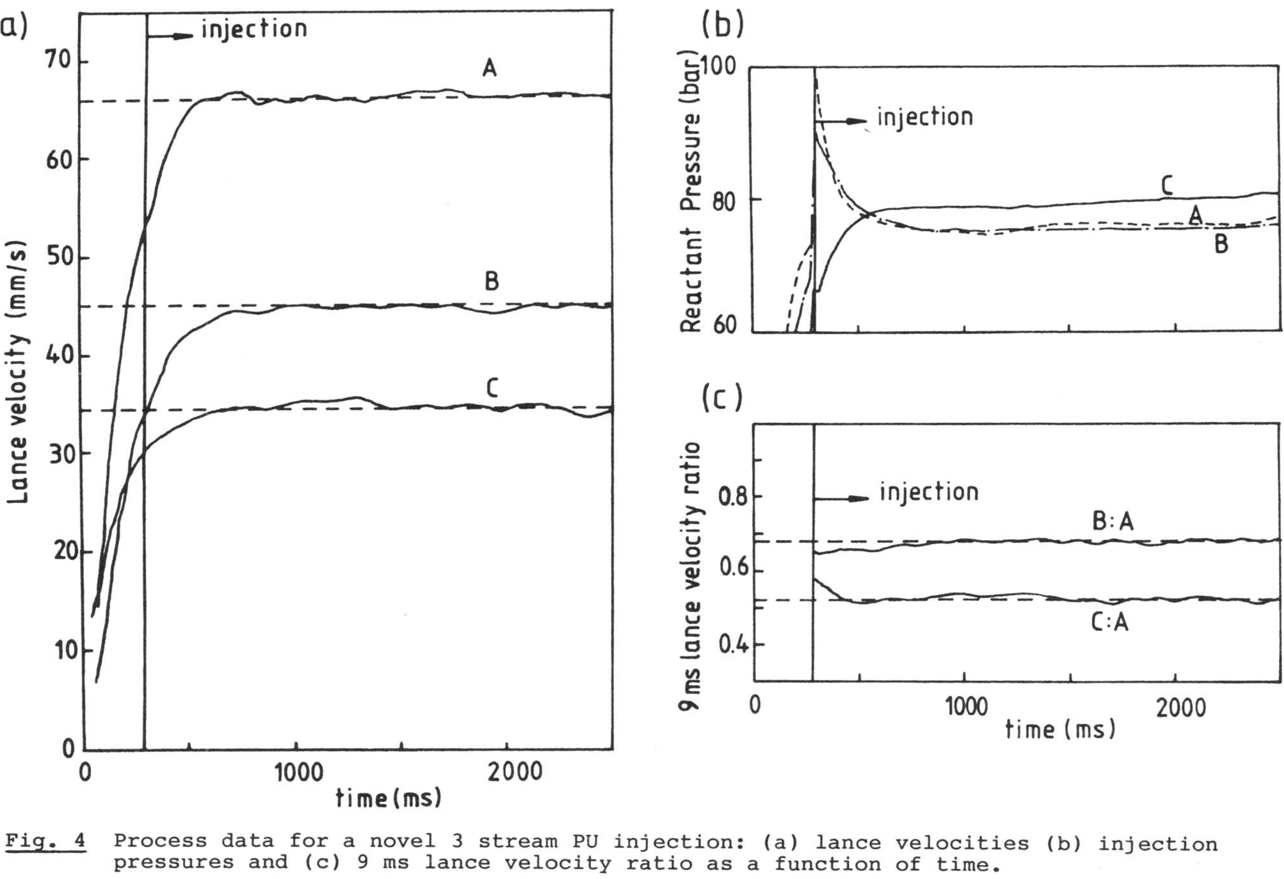

<u>Fig. 4</u>   Process data for a novel 3 stream PU injection: (a) lance velocities (b) injection pressures and (c) 9 ms lance velocity ratio as a function of time.

Part IV

Summation

MODERN SYSTEMS OF QC IN RELATION TO
MANUFACTURING PROCESSES

P.J. Roddy*

In order to prosper in todays economic climate industry must be committed to never ending improvement in products and services.
One of the tools which, in our view, is vital to achieve this on-going improvement is Statistical Process Control. The fact that it is a tool and not and end in itself is explained and that it must have involvement at all levels of the company in order to succeed.
The industrial environment in which we hope that recovery can be achieved by means of SPC are discussed and how we see SPC involving the production worker more with the ideals of the company.

I want to talk today about change. A change in the way we all run our business. I say our business because we are all part of it, each of our companies is part of a greater whole. The whole that I mean for the present purpose is western as opposed to Japanese industry. We in the west have been losing out progressively to the Japanese in the world market and more recently in our domestic markets.

We have been losing market share, losing the economic war because our products and services have failed to meet the expectations of our customers. Our principle downfall has not been only on price, it has in general been quality which has let us down in the eyes of our customers.

I expect that in the minds of many of you there will be a resentful, frustrated feeling generated at the previous statement on quality. I too have been frustrated by statements about "poor quality" when running a production function. Despite these frustrations we must, I think, recognise that we are competing with an industry which makes high quality products, high reliability products, in sectors where our industries cannot compete at present. We must accept that they, the Japanese, have a more effective way of running their manufacturing enterprises than we have yet found. Once we have accepted that, it frees us to seek the most significant differences in the way they do business and to find those which would be of benefit to us in our industrial environment. We must accept that we cannot, even if we wanted to, reproduce the Japanese national ethic in our society. We work within our own culture and are part of it. There are, however, Japanese working techniques which are generally used and which we in Ford are introducing both in our own factories and those of our suppliers. One body of these techniques collectively we call Statistical Process Control (SPC).

* Ford Motor Company Limited

The concepts and methods of SPC were, vexing as it may be, taken to Japan from America, as part of the US industrial restructuring effort, by Dr. J Edwards Deming. The ideas, first developed by Dr. Shewart in the Bell Laboratories in the late twenties, were generally ignored by world industry which was euphorically pursuing the ideas of mass production in the wake of the new deal. Our own British Standard was written in 1935 and ignored by most people until it was superceded by the development of the concepts which we now advance.

One fundamental difference is that the older B.S. approach is specification orientated, where the control limits were developed from externally imposed criteria. The SPC system is process related in that the control limits are derived from the output of the process.

This enables us to understand what the process "can" do rather than what it "should" do. If quality improvements are necessary then process change is called for.

Thus quality can cease to be a subject for argument and can by defining quality problems in terms of process capability produce improvement in process control and real quality through involving all the people in manufacture in a positive way.

The message which the Japanese accepted from Dr. Deming was -

. You can make quality and here is the way to do it - SPC.

. You have to do it to eat, you can send quality out and bring food back.

. Work with your vendors.

. Look to the future - produce goods which will have a market years from now.

. Redesign your products and processes to meet customer aspirations.

. Work on instrumentation - measure your outputs.

. Bring the process under control to achieve ever improving quality.

We are all familiar with the results of that acceptance. Japanese industry has accelerated away leaving us all wallowing in its wake.

We have effectively reversed the roles of our two nations in terms of "cheap and cheerful" (or nasty) versus "you have to pay for it but its good quality". We have come from a situation where "made in Britain" was a prestige lable to one where "made in Japan" carries the same cachet.

Can we afford to ignore the message which they accepted?

Not if we want to get back into contention.

So if Japanese quality is better than ours what do they mean by quality?

Quality is uniformity, the more uniform the product the better the inherent quality. Reducing variability is what quality improvement is about.

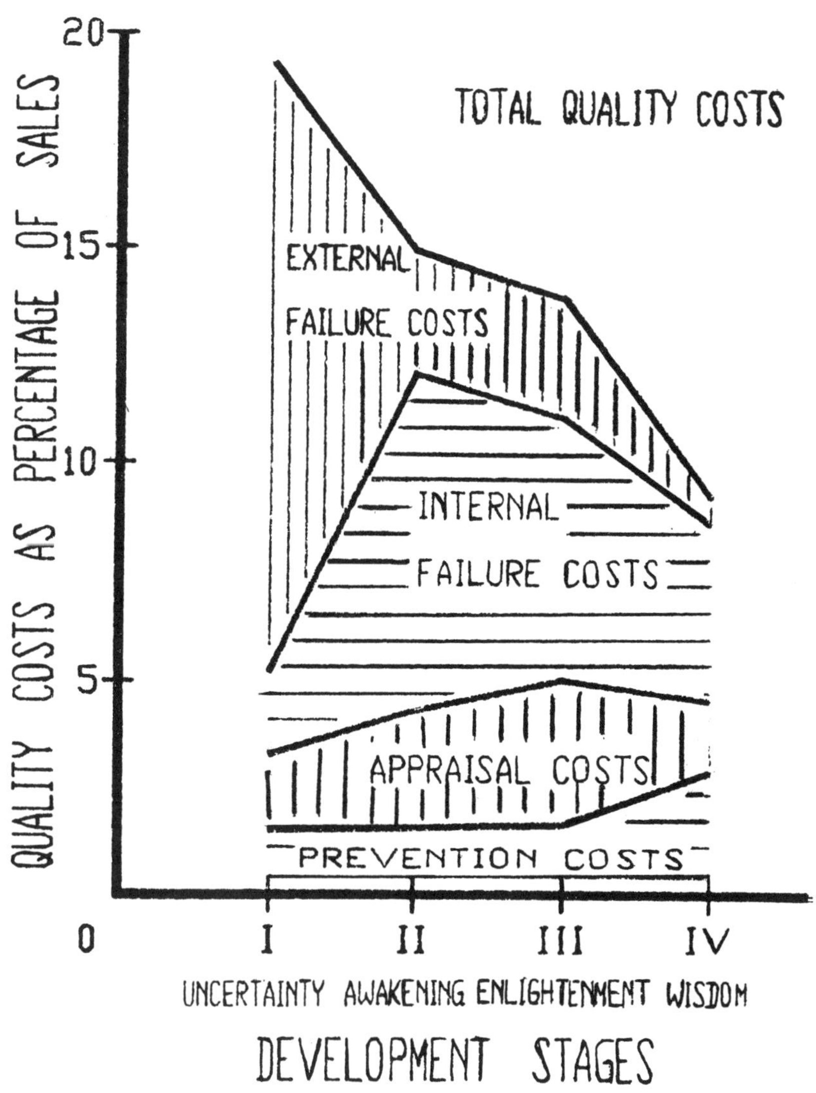

FIGURE (1) Total Quality Costs

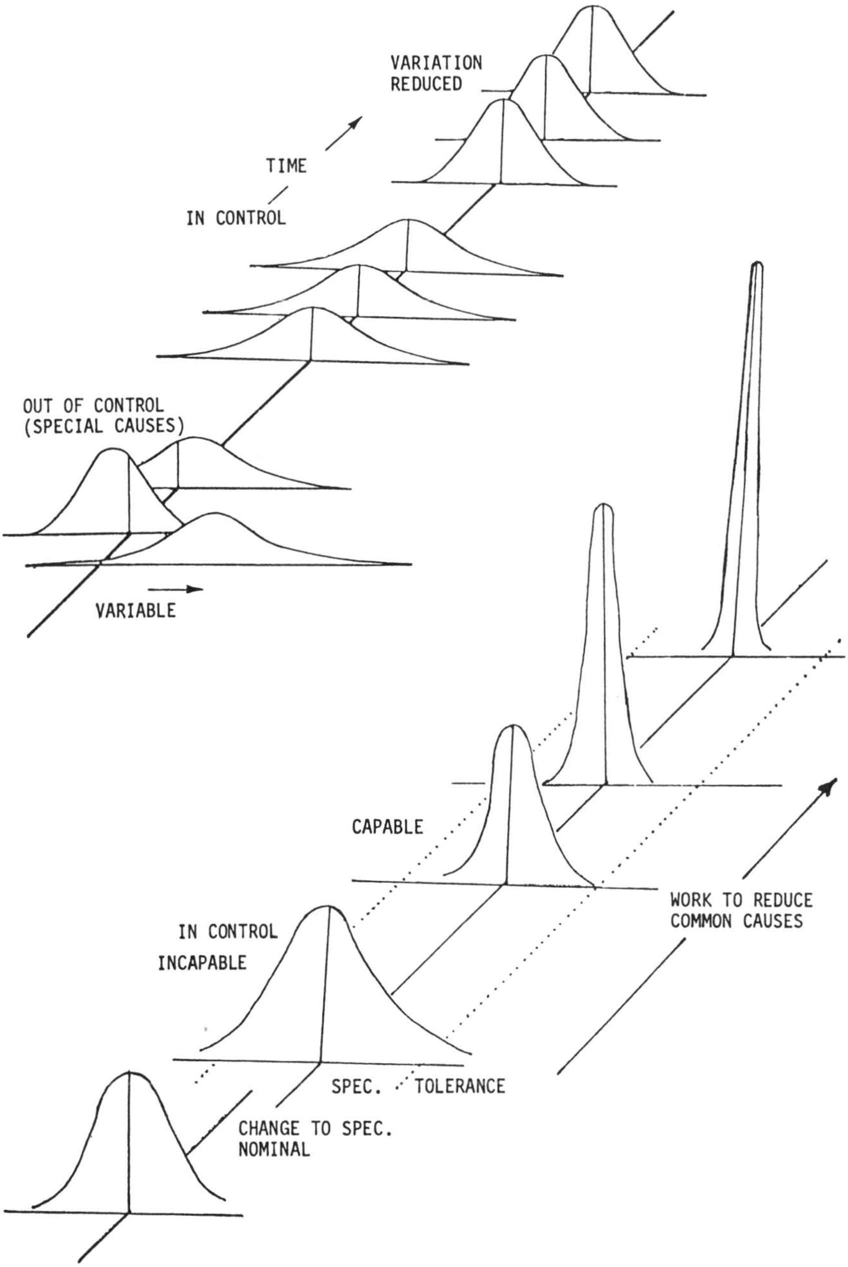

FIGURE (2) Development of Variability Reduction

The Japanese product is also cheaper, cheaper and better! This exposes one of our common misconceptions that quality costs money. Quality <u>saves</u> money. How many of your companies know, really know, what their <u>quality</u> costs are?

The chart (1) illustrates the change in costs of a typical company progressing from uncertainty to wisdom, it is schematic but not unrealistic.

You may look on 15% of sales as high for external failure but what is in there?

. Scrap.
. Rework at customers.
. Sorting at customers.
. Transport, both ways for no cash return.
. Quality discounts.
. Lost sales.
. Lost reputation.
. Price reductions to compensate for reputation.

Maybe 15% can be an understatement. So let us look at where we start from and see quality as an opportunity. No not a problem, an opportunity because if a company can survive, however marginally, with the situation portrayed think what it can do by tackling and rectifying its management of the process to achieve the condition which is portrayed as wisdom.

The changes in output can be traced in Figure (2). They show the evolution from out of control distributions to achieving statistical control and the beginning of the reduction in variability which must be permanently built into our thinking.

So how much do we need to improve?

This question, so often asked, shows a fundamental misunderstanding. Improvement in quality has to be a way of life not an event. The expectations of our customers will continue to increase in terms of quality, as we get better they will want to have ever better products. Our competition has had years of advantage over us in the pursuit of quality improvement, we have a long way to go. They are continuing their improvement programme. We have a moving target and must accept that there can be no end to the competitive improvement we must maintain.

This is the single most important idea I want to project today. We must maintain continuous improvement in products and services or go the way of the motocycle, optical and consumer electronics industries.

They all expressed the view that they had "good" products "good" markets. Where have they gone?

We in Ford do not intend to go down that road, I don't think you want to either. One method we are employing to avoid this is SPC.

Unfortunately there is not sufficient time to develop an understanding of the techniques of SPC today. SPC is however a body of techniques which enables the user to -
. appreciate what his process can produce in terms of variability,
. analyse the causes of variability,
. rank the causes to prioritise actions.

It is a discipline which maximises the significance of existing data and helps to identify areas in which data can provide improved understanding and control.

You do not have to be a mathematician to use it anymore than you need to be an automotive engineer to drive a car.

What can SPC do for your company? It can help to ensure its survival in a competitive environment.

The traditional approach to quality is to depend on production to make the product and on quality control to inspect it and screen out items not meeting specifications by actions on the product. Such as sorting, reworking or scrapping.

This involves a strategy of detection. It is wasteful because it allows time and materials to be invested in products or services that are not always usable. "After the event" inspection is uneconomical, it is expensive and unreliable and the wasteful production has already occurred.

It is much more effective to avoid waste by producing only usable output in the first place:- a strategy of prevention. SPC is about prevention by control of and action on the manufacturing process.

The existance of a department called "Quality Control" directs the minds of management to believe that the department is responsible for quality. It is not. The producer and designer are responsible for the quality of the goods. The existance of a "Quality Department" removes attention from this.

Let us look at a process control system - Figure (3).

A process control system can be described as a feedback system. Four elements of that system are important to the discussions that will follow:

1. The Process - By the process, we mean the whole combination of people, machines and equipment, input materials, methods and shop environment that work together to produce output. The way the total process has been designed and built, and on the way it is operated. The rest of the process control system is useful only if it contributes to improved performance of the process.

2. Information About Performance - Much information about the actual performance of the process can be learned by studying the process output. In a broad sense, process output includes not only the products, but also any intermediate "outputs" that describe the operating state of the process, such as temperatures, cycle times, etc. If this information is gathered and interpreted correctly, it can show whether action is necessary to correct the process or just-produced output. If timely and appropriate actions are not taken, however, any information-gathering effort is wasted.

3. <u>Action on the Process</u> - Action on the process is <u>future and change oriented</u>, as it is taken when necessary to <u>prevent</u> the deterioration of the process. This action might consist of changes in the operations eg. operator training, changes to the incoming materials, or the design of the process as a whole. The effect of actions should be monitored, and further analysis and actions should be taken if necessary.

4. <u>Action on the Output</u> - Action on the output is <u>past and acceptance oriented</u>, because it involves <u>detecting</u> out-of-specification products already produced. If current output does not consistently meet specification it will be necessary to sort all products and to scrap or rework any nonconforming items. This must continue until the necessary corrective action on the process has been taken and verified, or until the product specifications have been changed.

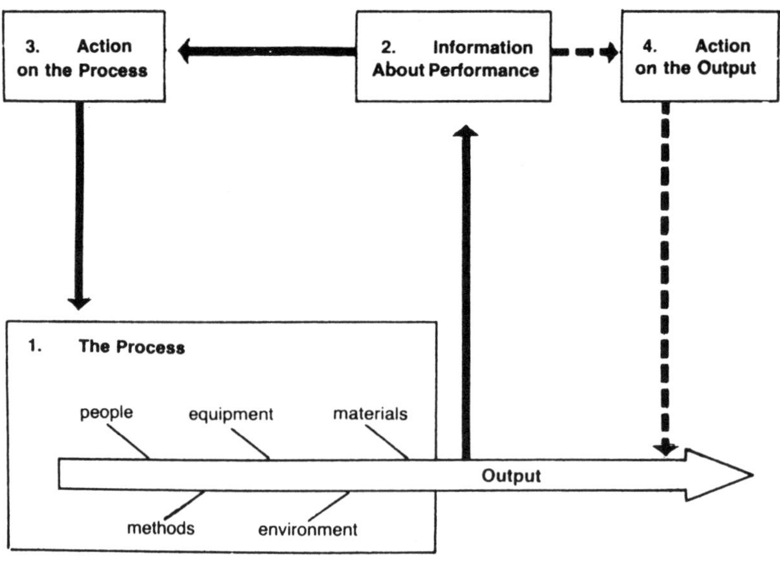

FIGURE (3) A Process Control System

What can Statistical Process Control do for this process? In order to establish a starting point for improving you have to be able to measure variation. Measuring each batch or part is the only way to obtain absolute precision but is impractical. Statistics enables us to take small amounts of data and use them to:

. Analyse special and common causes of variability to enable control to be achieved and improved.

. Identify by means of simple, operator applied charts, the ideal time to make process adjustments which prevent off specification material or components being made.

. Define the capability of the process by superimposing specification requirements onto process generated performance characteristics. Thus moving "quality" from the argument mode into the agreement mode because it is numerically or graphically defined.

Charting should not be seen as a cure all, it is a technique to effectively analyse data to uprate the quality of information which data conveys to the user.

SPC is not charting alone. To emphasise this point let me introduce the 20:10:70 rule. In SPC twenty percent of the effort goes into setting up and operating control charts, ten percent is just common sense, seventy percent of the effort goes to process improvements.

Charts alone will not produce the benefits. Work done on measuring and improving the uniformity of process outputs, as illustrated by charts will produce those benefits.

The application of SPC is intended to improve your business performance in all sectors. Techniques to establish you on the improving trail are detailed in the Ford SPC manual.

Be cautious SPC is powerful medicine, we do not recommend massive doses, start small and start simple! Take one area of your operation, train a small number of people who are to be concerned and allow them to start slowly.

If your organisation is to benefit overall from the application of SPC it follows that you must establish an environment suitable for action. Any statistical method will fail unless management prepares a responsive environment. Fear within the organisation which inhibits people from being candid must be removed. You will probably discover some unpleasant facts once you start. About half the processes investigated in UK have been incapable. If your investigating team (one man team perhaps) discovers that your process is not capable, he and the people on and around the job must be secure enough to come and tell you. The "kill the messenger" syndrome is common, the "who is to blame?" question still troubles our industry. If we are to become effective we must change these responses to daily problems. Produce an environment where solving problems is more profitable for an employee than making excuses, that makes the whole operation more profitable more quickly.

. Train people to do their jobs better.

. Train yourself to do your job better.

. Push supportive "no blame" management style down to grassroots.

. Involve operators in the control of their own process.

We have heard alot about automation and high technology this week. I would suggest that we are under-utilising the most complex factors in our plants. People!

We make things through people. We offer service through people. What can we do to make these people, that workforce, committed to achieve the improvement in quality which is necessary for our survival. We see SPC as an approach to running our business which makes this possible.

Many of our competitors' plants are not highly automated. Many of them in fact achieve excellent results from facilities which we consider dated. They have found a cheaper way to make more and better products and it works over here too.

In a factory which I visited recently which makes polymer based products I saw an increase in precision, productivity and worker morale which was projected to cost about a quarter of a million pounds in terms of machines. This improvement cost about two thousand pounds and the effort to involve the operators in an SPC programme.

One operator, the difficult one, said when asked about SPC.

"I don't get paid anymore for this you know", pointing to his chart "But I don't get shouted at about quality. We don't make any scrap because we can see it coming, so we would deserve a shouting at."

Talking to another operator I asked him what his specification was. His reply was "We don't pay too much attention to specification. It used to be plus and minus a millimetre but we can now control the machine to 0.3mm so that is what we do. We were kidding ourselves about the millimetre though we found that out once we started SPC."

It sounds almost like quotes from Japan but it is Britain.

So briefly let me review what I have said.
. We must change to survive in the face of competition.

. That change must include a commitment to continuous improvement in all our products and services.

. Quality can be defined in terms which reduce discord and produce positive motivation which will benefit general motivation.

. There are techniques for measuring improvement.

. Improvement should be based on defect prevention.

. Whilst long term benefits are the main aims of SPC the short term increase in costs can be rapidly recovered.

. There are effective routes to improve productivity and profitability which do not necessitate major reinvestment.

THE NEED FOR NEW TECHNOLOGIES AND ATTITUDES IN UK MANUFACTURING INDUSTRY

J.H. Smith*

<u>Summary of the final keynote address</u>

1 The pursuit of profit and viability in a
  rapidly changing world.

2 The internationalization of markets in a global
  economy and the rise of multi-national
  corporations.

3 The velocity of technological change and
  shorter product life cycles.

4 The enabling technologies.

5 Technology and the manager:

  - the pervasive influence of technology

  - a manager's understanding of technology

  - technology, economics and operations
    management.

6 The need for changes in attitudes and
  acceptance of high risk decision making in
  investment and human resources.

* Managing Director, Birkbys Plastics Limited, Liversedge, UK

BIOGRAPHICAL NOTES

Dr Michael Abrahams is Vice President of Product Support for General Electric Calma in Europe. Dr Abrahams graduated from the University of Strathclyde with a degree in Mechanical Engineering and a PhD in Bio-Engineering; he is also a member of the Institution of Mechanical Engineers.

Sydney Ashmall after taking an M.A. in physics and mathematics and an M.Sc. in pure and applied mathemtics, joined the National Engineering Laboratory at East Kilbride, which is the government's engineering research establishment. Apart from three years which he spent with the systems programming team at NEL, he has always worked on the application of computers to manufacturing. For the last twelve years he has been concerned with computer control of machine tools and has produced control communications systems which link them, and has produced systems for several industrial companies. He is now involved in robotics and in establishing general strategies for automation.

Roger Austin started out as an apprentice engineer and has been in automation since the age of 21 years. In 1976 he formed his company IAS Plasguard manufacturing conveyors and safety guarding. In 1979 he formed Pressflow Limited manufacturing pick and place units for the Injection Moulding industry. In 1982 he sold his first robot into Europe and in 1985 he ventured into the USA. 1984 saw Pressflow take the lion's share of the UK market of robots for injection moulding machines. In 1985 both Pressflow and IAS Plasguard became subsidiaries of the Thomas Dudley Group of Companies, Dudley, West Midlands, with Roger Austin retaining overall control.

Ian T Barrie is a graduate in Physics (London University) and has been a consultant to the plastics industry for the past 3 years. Prior to that he had 30 years of multi-product experience at ICI on the research and development side. About half this time was concentrated on injection moulding, new processes, new materials and applications; the remainder was divided equally between the study of mechanical and electrical properties. His interest in computers has been continuous since writing his first programs in 1963.

Dr G Bolder studied Electrical Engineering at the Technical University of Aachen, receiving a degree in 1980. For the following four years he was a scientific coworker at the IKV studying measuring techniques and control of the extrusion process. Since 1984 he has been the Chief Engineer of the IKV.

Paul J Clarke joined Bucher-Guyer (UK) Limited in 1985 as General Manager, being responsible for the UK Sales and Servicing organisation. Having started his working life at BIP Engineering, he spent several years in the development department working on control systems for Injection Moulding Machines. He subsequently transferred to the Sales Department for 6 years before joining Battenfeld as Product Manager.

P D Coates BSc MSc PhD   is senior Lecturer in the School of
Mechanical Engineering at the University of Bradford. After a
background of research in solid phase forming of polymers he has
more recently pursued wider interests in polymer processing,
including industrially collaborative research in the areas of
microprocessor/computer control of a variety of polymer processing
techniques, including reactive processing methods. Together with
Dr Johnson, he is responsible for a major research programme in
RRIM (supported by an industrial consortium and SERC) which has
involved the development of two novel design computer controlled
RRIM machines for a variety of chemical systems.

Dr Thomas Coppetti   was born in 1942, studied mechanical
engineering at the ETH in Zurich. After his studies there, he was
assistant-instructor and later on became a staff member at the
scientific institutes of regulation, steam sytems and machine
tools. In 1975 he was promoted Dr Sc Tech with his dissertation
on the stability of hydraulically regulated circuits. Since 1974
Dr Coppetti has been a managing partner at Netstal Machinery Ltd
Nafels/Switzerland, as the head of the technical department.

J Czerski   was educated at Leeds and Aston Universities. Prior to
joining BASF UK Ltd in 1976 he was employed at the Shell Carrington
Plastics Laboratory where he was involved in the development of new
methods of processing expandable polymers and W.C.B Plastics Ltd as
Production Engineering Manager. He is now responsible for market
development and technical service within the BASF UK Business
area. His main personal interest being product developments,
processing control and education in the Plastics Industry. J
Czerski has lectured frequently at various PRI events and is a
present member of the Manchester Section PRI main committee. He
obtained his FPRI in 1979.

Robin Enderby   is Managing Director of Blow Moulding Controls, who
manufacture control systems for plastic moulding machines, and are
agents for Moog Systems. He studied computer science at Trent
Polytechnic.

Michael I Iddon's   career commenced in 1956, initially by spending
some years devoted to the manufacture, assembly and installation of
machinery for the rubber and plastics industry in order to attain a
good understanding of the practical side of the industry. After a
similar period involved in Drawing Office design work, special
areas of interest developed. External qualifications were gained
including membership of the Institute of Chartered Engineers. The
foundations of the successful Iddon range of Extruders and L.C.M
vulcanising systems were his early responsibilities, Iddon Brothers
being the first company to commercially develop and market the
L.C.M salt bath system around 1963. A reasonably substantial part
of his working life is devoted to the requirements of running a
Company, since becoming Chairman and Joint Managing Director in
1971. Apart from this demand, M I Iddon actively pursues the aims
of continual R & D into new processing systems, updating of the
Iddon range of equipment, and responsiblity for world wide
travelling to keep in constant contact with Iddon's customers
wherever they may be.

Ron Kallus   is President of Eurometrics, a company engaged in
applying CIM to the plastics industry. Mr Kallus qualified as a
BSc in Electronics at the Technion in Israel. He has also
undertaken graduate work in computer science at MIT, USA. From
1975-8 he was involved in developing the GPS navigation systems

for the US Air Force, and special equipment for the Space Shuttle.
From 1978-82 he developed the 'Focus' production monitoring system
which has been sold only in the US Market. From 1982-4 he managed
the Intermetrics activity in Europe, transferring the 'Focus'
product to Europe, before becoming President of Eurometrics in 1984.

J Macintosh is Industry Marketing Manager for General Electric
Calma in Europe and has specific responsibilities for the Plastics
and Automotive Idustries. Mr Macintosh graduated from the
University of London with a degree in Mathematics and is a member
of the Institue of Mathematics and its Applications.

V J Osola CBE FEng FIMechE is Chairman of Advanced Manufacturing
Systems Committee of National Economic Development Office;
Executive Secretary, Fellowship of Engineering, London; Chairman,
John Osola and Associates Limited, Worcester. Mr Osola's career in
manufacturing industry included: gas turbine research, heavy
engineering, 15 years in the chemical industry, 15 years in the
glass industry, and 5 years in capital goods manufacture. He is a
former Chairman of the Mechanical Engineering and Machine Tools R &
D Requirements Board, Department of Trade and industry.

MacRobert Award Medallist, 1978; President of The Institution of
Mechanical Engineers, 1982: Fellow of the American Society of
Mechanical Engineers; Member of the Institute of Energy; a Fellow
of the Royal Society of Arts; a Freeman of the City of London. Mr
Osola's principal interest throughout his life in manufacturing
industry has been innovative development of production and
manufacturing processes.

M J Owen has worked in the Department of Mechanical Engineering
at the University of Nottingham for 29 years, where he is now
Professor of Reinforced Plastics. For the last 22 years he has
been involved in research on reinforced plastics and has a large
number of publications. Originally the work concentrated on a
wide range of fatigue properties of FRP. In more recent years the
research programme has been collaborative with various industrial
partners including the Ford Motor Co Ltd, Pilkington Brothers plc,
BTR Permali Ltd, ERF Plastics Ltd, and the Polymer Engineering
Directorate. Professor Owen is a Chartered Engineer and a Fellow
of the Institution of Mechanical Engineers and the Plastics and
Rubber Institute.

I Packman after serving an apprenticeship with a large British
electronics Company, joined an office equipment manufacturer as the
assistant to the Production Director. During the subsequent
decade, he took on the functional roles of Quality Control Manager
and Production Control Manager at two separate manufacturing
sites. In the latter post, he was responsible for the
implementation of the IBM COPICS system at the Company's major
European plant. He then filled the role of Materials Manager for
an STC Company where he successfully implemented MAAPICS on an IBM
S/34 minicomputer. Since joining Thorn EMI Datasolve, he has added
significant data processing experience to his engineering and
materials management background.

Dr John Parnaby is Group Director of Manufacturing Technology of
Lucas Industries and a Director of Joseph Lucas Ltd. He joined
Lucas in 1983 from the Diversified Products Sector of the Dunlop
Group where for 18 months he was General Manager-Engineering. His
role in Dunlop covered the design and installation of manufacturing

systems, automated machine systems, CAD, CAM, process and product development as well as the sale of these outside Dunlop.

His previous appointments include periods as Joint Managing Director of Rieter Scragg Ltd, University Professor of Manufacturing Systems Engineering, Works Director of Solway Chemicals Ltd, and a member of the Management Committee of the Marchon Division of Albright & Wilson Ltd. Dr Parnaby is a member of a number of government science and engineering committees and is a senior member of several professional and industrial bodies. He is a university graduate course external examiner and a visiting Professor of Manufacturing Systems Engineering.

Dr Parnaby is a Chartered Engineer with a first class honours degree in Mechanical Engineering from Durham University and a PhD in control Engineering from Glasgow University. He is a Fellow of the Institutions of Mechanical and Production Engineers, a Member of the Institution of Electrical Engineers, and a member of the Council of the Plastics and Rubber Institute.

P Roddy worked in the Steel Industry in both laboratory and production management before coming to Ford, working in laboratory supervision, materials engineering, management of laboratories and production. P Roddy is currently working for the director of product quality covering chemical and metallurgical matters for European manufacturing.

Bertil E Rosvall MSc is a manager of CAED at General Electric Plastics in The Netherlands he holds a masters degree in systems engineering from the Institute of Technology, Stockholm, Sweden. On completing his studies in Sweden he joined General Electric Plastics in The Netherlands, where the current CAED was being installed. He currently manages the unit in The Netherlands which provides analysis for customers support and market development activities as well as systems management for the CAED Network.

John H Smith was educated at Beal grammar school in Ilford, Essex, and joined the Plessey Company in 1951 as an apprentice in Jig and Tool Design and during the 1960's rose from a position of Production Manager to Divisional Manager in Company's Aircraft Electrical and Mechanical businesses. In 1970 he took over the Enfield Tool Company, a subsidiary of Plessey and developed a group of businesses known as Plessey Precision. In 1975, he left Plessey and went to South Africa. He was Managing Director of Pandrol (SA). In 1978 he was approached to rejoin Plessey and took over as Managing Director of Birkbys Plastics in September 1978 - the position he holds today.

B W Small has over 25 years experience in manufacturing industry helping improve manufacturing profitability and effectiveness. From Britain and the United States he has worked internationally. After graduation, completed in parallel with an apprenticeship at Westinghouse, he moved to an Engineering Consultancy in Bristol where he became involved in a broad range of projects including research into machine tools, design of special production equipment, factory layout, product reliability, and defence systems. This included engineered products, consumer products and consumer durables. His initial grounding with Ingersoll Engineers was in development of advanced machining and assembly operations and also in the development of the business criteria of why, how and when a company should improve its manufacturing performance.

<u>Mike Tee</u> is Chairman of the conference organizing committee. His 25
year association with the polymer processing sector and the PRI
started with attendance on the full-time postgraduate course in
plastics technology and polymer chemistry at the National School
of Rubber Technology (now the London School of Polymer Technology).
Service with Reed International, Ford, TRW-Clifford, Fothergill
and Harvey and Knight Wendling has provided exposure to most, if
not all, of the industry's processes and materials. An initial
career emphasis on research and development management changed to
a general management interest in the early 1970's, with operational
experience at board level being gained with both UK and US companies.

Since 1982 Mike has been responsible for developing Knight Wendling's
business activities in the polymer-based industries and a
professional interest in business administration and the implementation of AMT in the sector continues to be stimulated by a part-time
research programme at the Manchester Business School. Mike is
currently a Fellow of the PRI, the British Institute of Management
and the Institution of Industrial Managers.

<u>Dr Gerry Trantina</u> is a Manager in computer aided application
development of the Plastics Business Group of General Electric
Company, USA. He received his doctorate from the University of
Illinois, USA, for his work in applied mechanics. He joined the
General Electric Company in 1973 at the Corporate Research
Headquarters in Schenektady, New York, where he worked on
structural analysis. In 1983 he joined the Plastics Business Group
as Manager of the Caed Group dedicated to work on engineering
thermoplastics materials.

<u>Clive Turner</u> is a qualified Management Accountant with industrial
experience in manufacturing as a Management Accountant and Computer
Manager. He has retail experience (5 years consultancy and
training), as well as experience being Finance Director (3 years in
the holiday industry) and for the last 4 years managing his own
business of management consultants specialising in the financial
implications of manufacturing and the provision of related training
programmes.

<u>N Weatherby</u> received an upper second degree from the University of
Bradford in Textile Science and Technology. After spending 1 year
in the wool industry he then moved to Marks and Spencer for a
further 3 years as a textile technologist specialising in all types
of woven film. From there he worked for Rieter Scragg, a company
manufacturing yarn texturizing machinery, for 2 years as part of a
research and development team. During this period he also attended
Manchester University part-time for which he received an MSc degree
for 2 years research in fibre physics. He moved to Nottingham in
May 1984 and currently holds the position of research assistant in
the Department of Mechanical Engineering whilst studying part-time
for a PhD in composites.

<u>Theodore Zavos BSc CEng MIProdE</u> was born in Nicosia, Cyprus, in
1953 and has been a permanent resident of the UK since 1968. He
attended St Bees School in Cumbria and graduated as a BSc in
Mechanical Engineering from Middlesex Polytechnic, having
specialized in Production Engineering.

He spent three years as a Planning Engineer with Roneo Vickers Ltd
and worked on machine shop and automation projects. In 1978 he
joined Chloride Automation Batteries Ltd and dealt with the
introduction of automation to their battery finishing process.

In 1982 Mr Zavos joined Elco Plastics Ltd and was part of the team which relocated the factory, setting up the Thorn EMI Ferguson Specialist Components Division. He introduced pick and place robots for moulding machines and set up a robotic spray shop.

Since January 1986, Mr Zavos has been Manufacturing Engineering Manager with the Central Engineering Services Division of Thorn EMI Ferguson, based at Enfield in Middlesex. Some of his main projects include: automatic insertion of electronic components to PCB's, robotic assembly of television chassis, and robotic stapling of a moulded fascia to television cabinets.

CHEMICAL ADDITIVES FOR THE PLASTICS INDUSTRY
Properties, Applications, Toxicologies

by

Radian Corporation

This book describes in detail the chemical additives used in the plastics industry. It analyzes the chemicals used as additives in polymer manufacturing and the processing of plastics, environmental releases of these chemicals, and possible occupational exposures to them.

The development and growth of the use of plastics since World War II has been phenomenal. The technology for the manufacture of polymers and plastic products has expanded into the most important of the chemical-based industries, producing new products for new uses at a remarkable rate. Plastics additives have played a critical and complex role in this growth. The purpose of this book is to put additives used in the manufacture of plastic products in perspective for environmental and health impact analysis.

The plastics additives are presented as major functional groups of chemicals, which are further subdivided into chemically-, functionally-, or physically-similar chemicals. An overview of each major functional group includes the properties and applications of the subclasses, their environmental impacts, and possible occupational exposures. Common worker exposure practices for each functional group are also presented.

A series of appendices details the physical and chemical properties and polymer application of each chemical within the functional groups; the industrial, commercial, and consumer uses and consumption volumes for each chemical; and data on toxicological and worker exposure concerns for each chemical.

CONTENTS

ISBN 0-8155-1114-0 (1987)

884 pages

WATER-SOLUBLE RESINS
An Industrial Guide

by

Ernest W. Flick

This book contains descriptions of almost 1300 water-soluble resins which are available to industry. Natural and synthetic resins, with numerous uses and applications, are included. The book will be of value to technical and managerial personnel involved in the final products made with these resins, as well as to the suppliers of the basic raw materials.

The data represent selections from manufacturers' descriptions made at no cost to nor influence from, the makers or distributors of these materials. Only the most recent information has been included. It is believed that all of the products listed here are currently available, which will be of utmost interest to readers concerned with product discontinuances.

The following information, as presented by the supplier, is listed for each product, as available: company name and product category, tradenames and product numbers, and product description including main features as described by the supplier.

Listed below is a condensed table of contents containing **section titles and selected subtitles.** Parenthetic numbers indicate the number of products covered in that category.

1. **CELLULOSE ETHERS (141)**
   Sodium Carboxymethylcellulose
   Methylhydroxyethylcellulose
   Hydroxypropylcellulose
   Hydroxyethylcellulose

2. **GELATINS AND COLLAGENS (12)**

3. **NATURAL GUMS (214)**
   Carrageenan
   Guar Gum
   Gum Arabic
   Gum Tragacanth
   Locust Bean Gum
   Algin
   Xanthan Gum

4. **RESIN DISPERSIONS, EMULSIONS AND SOLUTIONS (780)**
   Vinyl Acetate-Ethylene Emulsions
   Polyvinyl Acetate Homopolymer Emulsions
   Polymer Deflocculants
   Polyacrylate Thickeners
   Acrylic Acid Polymers
   Water Reducible Resins
   Water Reducible Alkyds
   Cationic Quaternary Ammonium Polymers
   Water Active Lubricant Basestocks
   Flocculants
   Polyelectrolytes
   Urethane Resins
   Coating Resins
   Polyester Resins
   Polyurethane Dispersions
   Opacifiers
   Plastic Pigments
   Coagulants
   Binders, Saturants, Laminants for
     Nonwovens and Saturated Papers
   Ethylene-Vinyl Acetate Emulsions
   Polymers for Adhesive Manufacturing
   Polymers for Aqueous Inks
   Polymers for Building Products
   Polymers for Industrial Coatings
   Polymers for the Textile Industry
   Water Treatment Additives
   Metal-Complexed Acrylic Emulsions
     for Floor Maintenance
   Water-Borne Acrylic Cement Modifiers
   Scale Inhibitors
   Soil Resistant Polymer
   Carboxylated Styrene Butadienes
   Styrene/Acrylic Copolymers

5. **RESINS (146)**
   Polyethylene Imines
   Polyacrylamides
   Polyvinyl Alcohols
   Polybutenes
   Pyrrolidone-Based Polymers
   Phenolic Resins

6. **SUPPLIERS' ADDRESSES**

RESIN INDEX

ISBN 0-8155-1107-8 (1986)

222 pages

POLYMER MANUFACTURING
Technology and Health Effects

by

Radian Corporation

This book is a detailed analysis of the polymer manufacturing industry. Elements of the analysis include an industry definition, raw materials, products and manufacturers, environmental impacts, and occupational health concerns. Manufacturing processes for 27 different polymers are covered. The book is a companion volume to Noyes' *Plastics Processing.*

Raw materials for plastics and resins are industrial organic chemicals used as monomers, or plasticizers and specialty chemicals used as additives to modify resin properties. Polyethylene, polyvinyl chloride, polypropylene, and polystyrene are but a few examples of high volume polymers manufactured in the U.S. which have new uses emerging constantly. The types of companies involved in polymer production vary, but the principal producers include major chemical, petroleum, paint, tire and rubber, steel, and electrical manufacturing companies.

Some of the chemicals used as raw materials in polymer production are highly toxic and may produce serious adverse health effects in overexposed employees. However, effective engineering controls and personal protective equipment and clothing exist that greatly reduce worker exposure potential. Successful application of these controls depends on plant-specific factors such as plant design, materials handled, process configuration, and management and employee dedication to maintaining a good occupational health program, all of which are discussed in this book.

CONTENTS

1. INDUSTRY ANALYSIS
2. ACRYLIC RESINS
3. ACRYLONITRILE-BUTADIENE-STYRENE (ABS)
4. ALKYD MOLDING RESINS
5. AMINO RESINS
6. MODIFIED POLYPHENYLENE OXIDE AND POLYPHENYLENE SULFIDE
7. EPOXY RESINS
8. FLUOROPOLYMERS
9. PHENOLIC RESINS
10. POLYACETAL
11. POLYAMIDE RESINS
12. POLYBUTYLENE
13. POLYCARBONATE
14. POLY(ESTER-IMIDE) AND POLY-(ETHER-IMIDE) RESINS
15. POLYESTER RESINS (SATURATED)
16. POLYESTER RESINS (UNSATURATED)
17. POLYETHYLENE—HIGH DENSITY (HDPE)
18. POLYETHYLENE—LINEAR LOW DENSITY (LLDPE)
19. POLYETHYLENE—LOW DENSITY (LDPE)
20. POLYPROPYLENE
21. POLYSTYRENE (GENERAL PURPOSE)
22. POLYSTYRENE (IMPACT)
23. POLYURETHANE FOAM
24. POLYVINYL ACETATE
25. POLYVINYL ALCOHOL
26. POLYVINYL CHLORIDE
27. POLYVINYLIDENE CHLORIDE (PVDC)
28. STYRENE-ACRYLONITRILE (SAN)
BIBLIOGRAPHY
APPENDIX A—TYPICAL PHYSICAL AND CHEMICAL PROPERTIES OF POLYMERS
APPENDIX B—INPUT MONOMERS
APPENDIX C—COMPANIES THAT PRODUCE PLASTICS

ISBN 0-8155-1090-X (1986)

718 pages